HERU-COPTERS

AFRICAN AERONAUTICAL ASCENSION

By

African Creation Energy

July 07, 2014

WWW.AFRICANCREATIONENERGY.COM

ISBN 978-1-312-33460-1

Printed in the United States of America

"…By decree of Heru, Lord of the sky…

I fly up from you oh mortals.

I am not for the Earth;

I am for the Sky.

I have soared to the sky as Heru,

I have kissed the sky as a Falcon.

I will traverse the sky forever…"

~ "I will Fly: An Ascension Text",
Pyramid Text Utterance 467

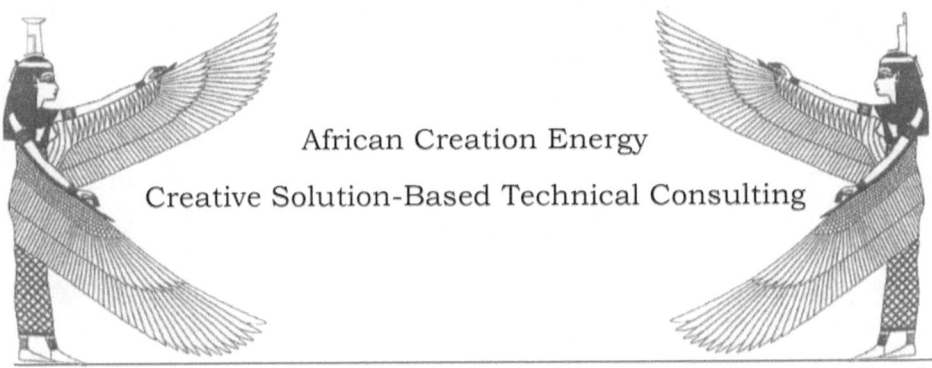

African Creation Energy

Creative Solution-Based Technical Consulting

TABLE OF CONTENTS

1. Ascension

"**Up**, **up**, *you mighty race, you can accomplish what you will*"; these are the **uplifting**, **revolutionary**, and **liberating** words spoken by the **Honorable Marcus Mosiah Garvey** in the early 1900's which expressed the sentiment that through work and effort it is possible to achieve the accomplishment of **Freeing** one's self from **repulsive** circumstances, and **Ascending** to the heights of your desires. Being able to "**Rise Up**" or "**Fly**" or "**Ascend**" has consistently been symbolic and representative of Freedom and Liberation throughout history: from the African-American folktale of "***The People Could Fly***" from St. Helena Island in South Carolina which told a story about slaves who had the ability to fly up out of bondage in America to freedom in Africa; to the Biblical story in the book of **Exodus chapter 19 verse 4** where **Jehovah** lifts the **children of Israel** out of bondage on **Eagle's wings**; to the Christian gospel hymn entitled "***I'll Fly Away***" which sings lyrics of removing the cold iron shackles of prison and flying away into freedom; to the uplifting poem entitled "**Still I Rise**" by **Maya Angelou** which uses flight, ascending, or rising, as a metaphor to represent transcending the horrors of slavery with the words:

> "...*Leaving behind nights of terror and fear,*
> *I rise into a daybreak that's wondrously clear.*
> *I rise bringing the gifts that my ancestors gave,*
> *I am the dream and the hope of the slave.*
> *I rise, I Rise, I RISE*"

With flight as a symbol of freedom, and the recognition that it takes work, effort, science, and creativity to fly, then we must also recognize that it takes a lot of work, effort, science, and creativity to obtain freedom.

The fact that both freedom and flight are obtained by way of work is expressed by a proverb from the **Akan** people of **Ghana West Africa** which states "*fawohodie ene obre na enam*" which is translated into English variously as **"Freedom comes with responsibilities"** or **"Liberation walks with struggle"**. The idea that "freedom is associated with struggle" is expressed in America in English by the idiom **"Freedom is**

Fawohodie – African Adinkra symbol for Freedom & Liberation

not free!" The West African Akan proverb is symbolized by an **Adinkra** symbol from West Africa called **Fawohodie** which represents **freedom, liberation, emancipation**, and **independence**. The relationship between flight and freedom can be seen in the Fawohodie Adinkra symbol which appears to be an abstraction of birds flying or butterfly wings. Considering the analogies which exist between freedom, liberation, independence, and Flight, perhaps there are analogies and lessons to be learned between **the scientific principles required to achieve flight**, and **the activities necessary to obtain Freedom and Liberation**. Since flight, freedom, and liberation require much work, the question must be asked, **"What is freedom and what does it mean to be free**?" When we speak of Liberation, what does it mean to be liberated; Liberation from what? Until you are able to fly, you are bound to the ground in pursuit of flight. Until freedom and liberation is achieved, you are a slave to your quest for freedom and liberation. To quote the poetic words of **Kahlil Gibran**: "*I have seen you prostrate yourself and worship your own freedom... I have seen the freest among you wear their freedom as a yoke and a handcuff... for you can only be free when you cease to speak of freedom as a goal and a fulfillment... In truth that which you call freedom is the strongest of these chains... And thus your freedom when it loses its fetters becomes itself the fetter of a greater freedom.*"

Simply put, "**Freedom**" is the **power to be able to do whatever you want to do without hindrance, restraint, or limitation**. Given the aforementioned definition of "freedom", it is necessary to realize that constraints exist on everything from the microcosmic to the macrocosmic, and therefore there is, and will always be, limits on freedom; just like there are limits on flight (the sky's the limit?). You are only free to do whatever you have been created **capable** to do, within the confines, constraints, and limits of the "**laws of Nature**" or the "**laws of Physics**"; this is **Natural Freedom**. Although the "laws of Nature" are the limits of Natural Freedom, it is Natural Freedom that is the form of freedom that is supposed to be "free" to all beings in Nature. Conversely, Natural Freedom could also be seen as a form of "**Natural Slavery**", in that all beings in Nature are **slaves to the laws of Nature**. It is the mastery of these laws of Physics (or laws of Nature), and using and applying one law to overcome another law, which gives one the sense of freedom; just like it is the mastery of the laws of Physics, and using the laws of thrust and lift to overcome the law of gravity, which achieves flight.

Using one law of Nature to overcome another law of nature, to create a sense of freedom, is **Artificial Freedom**. Economic systems, financial systems, and political systems have all been created by human beings to achieve a sense of artificial freedom from Nature. Ultimately, the primary beneficiary of the sense of "freedom" created by these artificial systems is the creator, or creators, of these artificial systems. All other people who participate in, or are a part of these artificial systems, are essentially slaves to the system, and this is **Artificial Slavery**. For example, a Financial and Economic system may be created to give people a sense of freedom from Nature where the people may not have to farm the land for their own food or obtain resources from nature to make their own clothes and build their own houses, but the people still have to work jobs to obtain money to purchase food, clothing, and shelter (the basic human needs) within the Financial and Economic system. So the people may feel they are free from having to work in Nature, but they still have to work in the artificial Financial and Economic system they are participating in to feel like they have "escaped" nature. Rather than being slaves to Nature, they become **"Debt Slaves"** to the artificial Financial and Economic system, and in time they will desire **"Financial Freedom"** from the artificial Financial and Economic system; and the cycle continues!

Birds, Bats, and Bees are naturally capable of Flight, and Human beings are not naturally capable of flight, but Human beings can create artificial ways to fly. So we recognize that being able to **CREATE** gives one the ability to obtain Freedom and Liberation. Science, Math, **Divergent Thinking**, and **Will** (Desire/Motivation) gives one the ability to CREATE. So when we speak of "Freedom," we are actually speaking of "freedom" comparative and relative to something else in Nature. The realization that freedom is a comparative and relative concept begs the question **"as 'free' as what"**; hence the metaphorical aphorism **"As Free as a Bird**!" From the Human perspective, being able to fly provides one with the freedom to travel wherever is desired in short spaces of time; flight is seen as a way to transcend the boundaries of space and time. Being able to fly provides a sense of freedom and liberation in that you seemingly have the ability to **travel** anywhere you would like to travel. One of the definitions of **"Liberation"** is **"the separation from adverse forces"**. An "adverse force" is anything that prevents you from obtaining any part of the **"hierarchy of basic human needs"**. The "hierarchy of basic Human needs" is what is needed for survival and well-being, and gives people the **Will (Desire/Motivation)** to take action. If you are being enslaved, but you are having the "hierarchy of basic Human needs" met, then you will likely lack the motivation and will to be free. Therefore, in order to be physically liberated, you must physically separate from anything you deem adverse, and in order to physically separate, you will have to **physically travel**, and move from one location and position to another. Thus, flight is not only symbolic of liberation, but flight as a mode of transportation, is also an actual means to physical liberation. The desire for liberation is what leads to the creation and consumption of transportation.

Being able to fly is symbolic of being able to "**do the impossible**"; you can do whatever you want to do. Being able to fly is both literally and metaphorically **Divergent**. Being able to fly affords one with the ability to transcend earthly boundaries which are seen as trivial, mundane, and superficial, and **ascend** to a "**higher realm**" in the **Heavens**. **Ascension** is similar to **Transcendence** and **Divergence**. "Flight" is also symbolic of "**Divergent Thinking**". Divergent Thinking is a thought process which considers many possibilities and is used to come up with creative ideas and solutions to problems. Divergent Thinking is often found amongst people who are non-conformist, who "think outside of the box", and consider concepts which are considered nonconventional. Esoteric, Metaphysical, Occult, Spiritual, Fantasy, and Paranormal topics are all examples of subject matter which can lead to Divergent Thinking, and hence, can lead to Creative ideas and new inventions. It is easy to see how flight can be seen as symbolic of Divergent thinking. If you were flying, and able to have a "**Bird's Eye View**" of the world (**the all-seeing eye**), you would immediately have a broader perspective which diverges, or is different from, the perspective and thoughts you would have by only ever having been on the ground. Flight provides you with the ability to ascend to the sky to obtain both a literal and metaphorical "**elevated perspective**", which produces a "**higher consciousness**".

The Sky is synonymous with "Heaven" or the "Spiritual Realm" and in various Religious and Spiritual traditions around the world, Heaven has always been the ultimate destination, therefore flying is a way to actually **"Ascend into Heaven"** to be with God who is called **"The Most High"**. This is why the **"Holy Spirit"** in Christianity is symbolized by a flying creature like the **Dove**. The association between flight, birds, Heaven, and the sky is also why **Angels** are often depicted with **wings**. Even in Africa in Ancient Egypt, aspects of the Soul such as the **Ba** and **Akh** (the unification of

Above: The Dove, Christian symbol for the Holy Spirit

Above: Ba – Egyptian symbol of the Soul

the **Ka and the Ba**) were depicted as birds or winged flying beings. And while the Abrahamic Religions of Judaism, Christianity, and Islam suggest that people have to die before they are allowed to **Ascend to Heaven**, birds, insects, and aircraft pilots ascend to Heaven, and experience Heaven daily in their life time without having to die. However, there are characters and rare situations in the Abrahamic Religions where people are believed to have **ascended into Heaven without dying**, and these types of **Ascensions** are viewed as **highly spiritual** and **mystical events**. In the Islamic religion, it is said that the **Prophet Muhammad** was able to Ascend to Heaven while he was alive by riding on a **Winged-Horse** called **Al- Buraq**. In the Bible, in the book of Kings II Chapter 2, Verse 11, it states that while the Prophet Elijah was alive, he was able to **Ascend to Heaven in a whirlwind on chariot of**

fire. In the Christian Bible, "**The Rapture**" is described in the book of 1 Thessalonians chapter 4 verse 17 where it states that people who are alive during the rapture will Ascend to heaven and be "**caught up in the clouds in the air**." And perhaps one of the most famous Ascension stories is the Christian religious story of the **Ascension of Jesus** into Heaven 40 days after his resurrection which is described in the Biblical New Testament in the book of Acts chapter 1 verse 9 through 11.

Scholars such as **Anthony Browder** and **Ashra Kwesi**, and documentary movies such as **Zeitgeist** and **Religulous**, have identified the similarities which exist between the Biblical Jesus character and the older African character of **Horus** from Ancient Egypt. Horus, whose name is believed to have been pronounced **Heru** or **Haru** in the Ancient Egyptian language, is an important deity from Ancient Egyptian cosmology and mythology who shares many similarities and characteristics to Jesus and other messiah or "savior-type" characters found in various religions. In his 1907 book entitled "**Ancient Egypt, The Light of the World**", the Egyptologist scholar **Gerald Massey** identified over 200 parallels between the Biblical Jesus and the Ancient Egyptian Horus including:

- Jesus was born without a father and Heru (Horus) was born without a father

- The Depiction of the "Madonna and Child" with Jesus being breast fed by Mary is predated by thousands of years by the Depiction of Heru (Horus) being breastfed by his mother Aset (Isis).

- Jesus is said to be the Son of God, and Heru (Horus) is said to be the Son of God (Ausar/Osiris) and also a Sun-God

Also, Heru (Horus) is related to Ascension in that Heru is often depicted as a falcon-headed man or as a falcon that ascends to the skies as it is written in the Ancient Egyptian Pyramid Text Utterance 467. Additionally, the name "Heru" is thought to have meant "**one who is above or over**", providing another relationship between **Heru** and **Ascension**. Moreover, as we previously discussed, Flight and Ascension is considered symbolic of Freedom, and Heru is not only related to flight and ascension, but Heru is also related to freedom and liberation. In Swahili, the word for "**Freedom**" is **UHURU**, and in Arabic, the word for "freedom" is pronounced "**Huria**" or "**Hurriyah**" (حرية). Both **UHURU** meaning "freedom" in Swahili and Hurriyah (as well as other Arabic words with the trilateral root H-R-Y) meaning "freedom" in Arabic, are etymologically related to the **Coptic** Egyptian word *Hōr*, which comes from the Ancient Egyptian

(**Medu Neter**) word Heru (Horus) . Therefore, one of the etymological meanings of Heru is "freedom". Other relationships which exist between Heru to concepts related to Ascension include:

- **Uhuru Peak** is the name for the **Highest point** of the **highest Mountain** in Africa (Mount Kilimanjaro)

- In Hebrew, "**Har**" הַר (Strong's H2022) means **mountain** or **Hill** (something "High-up" or Ascended)

- The Islamic Prophet Muhammad **Ascended** up a **mountain** to a cave named **Hira** to receive his first revelation of the Quran

- The name "**Harun**" in Arabic means "**Superior**" (something up-high)

- The word "**Hiero**" as in "Hieroglypics" means "something holy, sacred, or **set apart** (up high or Ascended)

From the exoteric or literal perspective, the numerous parallels between Heru and Jesus have led many people to conclude that Jesus was not a real person and that the story of Jesus was actually taken from the older and more ancient story of Heru. From the esoteric or symbolic perspective, there is recognition that both the story of Jesus and the story of Heru serve as allegories which are symbolic representations of a certain "**state of mind**" or type of **consciousness**. The esoteric understanding of the stories of Jesus or Horus has given rise to the **New Age** terms of "**Ascended Masters**" and "**Christ Consciousness**" or "**Krishna Consciousness**" which is promoted amongst the **Hare Krishna** religious group who use a combination of the words Horus (Hari) and Christ (Krishna) as the name of their organization, and chant the words "*Hari Krishna, Hari Krishna, Krishna Krishna, Hari Hari, Hari Rama, Hari Rama, Rama Rama, Hari Hari*" as a mantra for Ascension. In the "Ascended Masters" teachings developed by **Helena Blavatsky** for the **Theosophical Society**, "Christ Consciousness" is considered a **higher** form of consciousness or a **higher** form of spiritual enlightenment that is believed to have been obtained by **highly** advanced spiritual adepts called "**Ascended Masters**" such as **Jesus** and **Buddha**. One of the symbols used by Helena's Blavatsky's Theosophical Society was the **swastika** or "**Black Sun**" symbol which she adopted from eastern esoteric and mystical traditions like **Hinduism**, **Buddhism**, and **Jainism**, and the symbol can also be found in the **Congo** region of Africa as the **Dikenga**. It was this region of Africa which was claimed by **Germany** in the late 1800s during the "**Scramble for Africa**". Later in the early 1900s, inspired by mystical and esoteric teachings, the **Nazi** party of Germany adopts the swastika as one of its official symbols. There are numerous "Mystical" or esoteric spiritual traditions around the world that utilize allegorical stories and metaphorical tales as ways to teach what they consider as

"Higher" spiritual knowledge. Consider this: **Herons** (the birds) get physically high by flying, while **Heroin** (the drug) gets you metaphorically "high" by putting you in a different state of mind. In Hinduism, Buddhism, and Jainism, the direction of the arms of the swastika symbol are used to indicate either and Exoteric (external) or Esoteric (internal) teaching. If the arms of the swastika are facing right, it represents an Exoteric or External teaching, and if the arms of the swastika are facing left, it represents an Esoteric or Internal teaching. Also, the left-facing and right-facing swastikas can be combined into a single symbol called a "Sun-Cross" or **"Sun-Wheel"** which represents a combination of Esoteric and Exoteric teachings. In Buddhism, the four arms of the swastika represent the **"Four Noble Truths"** which are: 1) suffering, 2) the origin of suffering, 3) the end of suffering, and 4) the path that leads to the end of suffering. The swastika symbol influenced the **"Sun Cross"** or **"Sun Wheel"** symbol which is used in the Esoteric sect of Christianity called **Gnosticism** (meaning **knowledge**). Even the Abrahamic Religions of Judaism and Islam each have a "mystical" sect, namely **Kabbalah-ism** and **Sufi-ism** respectively, which acknowledge the esoteric or inner allegorical meaning to the stories presented within the religion as a path to "Ascend" to higher consciousness. In Kabbalah-ism the four-types of interpretations are 1) Peshat-the direct meaning, 2) Remez-the allegorical meaning, 3) Derash-the comparative meaning, and 4) Sod-the secret meaning. In Sufi-ism, the four main points are 1) Sharia-the moral code, 2) Tariqa-the esoteric path, 3) Haqiqa-the ultimate mystical truth, and 4) Marifa-mystical intuitive knowledge. In the allegories of most of the mystical and esoteric traditions, the Earth or the Terrestrial realm is representative of the physical, tangible, empirical world, and the Sky or the Celestial realm is representative of the mental, abstract, rational world.

Levels of Ascension in Various Mystical Traditions			
Buddhism 4 Noble Truths	**Judaism Kabbalah-ism**	**Islam Sufi-ism**	**Christian Gnosticism**
1) the path that leads to the end of suffering	1) Sod-the secret meaning	1) Haqiqa-the ultimate mystical truth	1) Gnosis - knowledge
2) the end of suffering	2) Derash-the comparative meaning	2) Marifa-mystical intuitive knowledge	2) Sophia - Wisdom
3) the origin of suffering	3) Remez-the allegorical meaning	3) Tariqa-the esoteric path	3) Nous – the demiurge, conscious intelligent mind
4) suffering	4) Peshat-the direct meaning	4) Sharia-the moral code	4) Logos – reason, the word

Interestingly, the esoteric and metaphysical information presented within the various Mystical Spiritual traditions which is said to lead to an "Ascended" higher spiritual consciousness, is also purportedly said to give the "Spiritual Adepts" the ability to physically "Ascend" in the form of **Levitation**. Simply put, the esoteric information leads to an exoteric manifestation. In Hinduism, Gurus who have ascended in their consciousness are said to have the ability to physically Levitate. Physical ascension in the form of Levitation is also anecdotally described in the mystical traditions of Buddhism, Gnosticism, and in the **Theosophy** of **Helena Blavatsky's Theosophical Society**. In the Sufi order of Islam, the Whirling Dervishes spin and rotate their bodies in repetitive circles (like a **Helicopter propeller**), which they believe gives the individual the ability to **levitate and Ascend** to become **Al-Qutb** (pronounced "Al Kubt"), a word used to describe the "perfected being". This Arabic word Qutb (ق طب) means the **axis**, center, or **pivot point**, and can be used to refer to the **central point of Revolution** on a **Helicopter propeller**. The Arabic word "**Qutb**" is also related to the Arabic word "**Qibla**" which is used to describe the Islamic direction for prayer called the **Kaaba** (**Kabah**).

Vehicles and Tools of Ascension from Various Cultures

Hindu Swastika

Dharma Wheel - Noble 8-fold Path

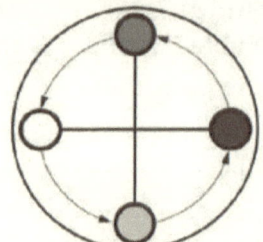

Congo Dikenga

Black Sun

Gnostic Cross

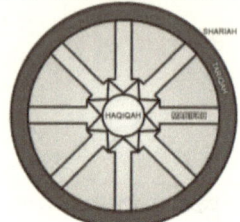

The 4 Spiritual Stations in Sufi Islam

Thor's Hammer fylfot Spinning

Ptah's Hammer Djed-Ankh-Waas

Native American Medicine Wheel

Utu, Shamesh, Nibiru- Suns/Stars in Mesopotamia

Roundel BMW symbol

Merkabah Ezekiel's Wheel within a wheel

In Hebrew and Jewish Mysticism Traditions like **Kabbalah**, the esoteric Ascension information is said to lead to an exoteric, or actual physical ascension, in a craft called a **Mer-Kabah**. During meditation, it is said that the body can be turned into a craft called a **Merkabah** which can be rode into ascension. The Hebrew word "Merkabah" (Strong's H4818) means "chariot" or "to ride", and is associated with the craft mentioned in the Hebraic Bible called **"Ezekiel's Wheel"**. If taken literally, the Hebraic Bible describes another craft used for physical ascension. In the Book of Ezekiel, Chapter 1, verses 1 through 28, a scene is described where a **whirlwind** comes out of the sky and brings forth a **wheel-within-a-wheel** which has **4 faces and 4 wings**. This "wheel-within-a-wheel" is called "Ezekiel's Wheel" and is called a **Merkabah** because the Hebrew word for wheel (Strong's H212) used in the Book of Ezekiel is also used to refer to the wheel of a chariot or Merkabah (Strong's H4818). The word used for "*wheels*" is **ophanim** (Strong's H212), which is also used in the **Dead Sea Scrolls** to refer to a group of Angels, the leader of whom is named **Razi'el** (meaning secrets of El [God]). In the Hindu Mysticism tradition, physical ascension is said to have been accomplished in crafts called **Vimanas**. In Ancient Hindu texts, the word **"Vimana"** is used to refer to chariots of the Gods that are able to traverse the sky. Depictions and descriptions of the Hindu Vimanas resemble **Ancient Egyptian Pyramids** which the Ancient Egyptians believed facilitated physical Ascension by sending the deceased Pharaoh to the celestial realm of the stars; this is why pyramids were built aligned to certain stars and celestial constellations. The Ancient Egyptian word for pyramid was **"Mer"**. Other depictions and descriptions of Hindu Vimanas resemble **"Sun Boats"** or **"Solar Chariots"** described and found in Ancient Egypt which were believed to be ships used to sail across the sky. The Qutb, Kaaba, Mer (Pyramid), Merkabah, and Vimana are all examples of crafts used for physical ascension as described by various mystical traditions.

Another exoteric manifestation of esoteric information is the fact that the airline called **EgyptAir**, uses the Ancient Egyptian symbol of Heru as the symbol for their company.

Amongst the mystical narratives from **Native American Indian** tribes, meditating and chanting in the center of "**Medicine Wheels**" (which were large circles constructed out of stone) would serve as target landing points for **Thunderbirds** to take people up on Ascension vision quests.

Inspired by the story of the Islamic Prophet Muhammad being able to ascend into heaven by riding the winged-horse called **Al-Buraq**, a **Moorish Aviator** from **Cordoba, Spain** named **Abbas Ibn Firnas** became arguably one of the first, if not the first, human in contemporary times to achieve flight in the 9th Century C.E. **Abbas Ibn Firnas** studied the flight capabilities of bats and birds, and designed, invented, and successfully flew a winged-glider **apparatus** for 3 miles by jumping off of a tower.

**Left:
Statue of
Moorish Aviator
Abbas Ibn Firnas
in Baghdad, Iraq**

Motivated by studying the esoteric information of various ancient mystical traditions, the Nazi party of Germany adopted the Swastika with the arms turned right (denoting an exoteric or physical interpretation) as their symbol. By studying the swastika and its associated doctrines, it is rumored that the Nazis developed disc-shaped, saucer-shaped, and bell-shaped physical crafts for ascension, and that all of the **U.F.O. (Unidentified Flying Object)** sightings seen on Earth are in fact crafts not from outer space, but crafts developed and build right here on Earth based on Nazi aircraft designs. Included in the "**Nazi U.F.O.**" rumors is the idea that in the early days of the **BMW** car manufacturing company, they produced several engines for aircraft, and some of these aircraft were also Nazi U.F.O. crafts, and this is why the BMW roundel logo looks so similar to the Black Sun symbol used by the Nazis. As will be discussed later in this book, "flying saucer" or disc-shaped U.F.O. crafts are designed and built on aerodynamic principles very similar to helicopters. British Science-Fiction writer Arthur C. Clarke is quoted as saying "*any sufficiently advanced technology is indistinguishable from magic*," and if the Nazi U.F.O. rumors are true, then we can also say that "*any sufficiently advanced technology is easily mistaken as extraterrestrial.*"

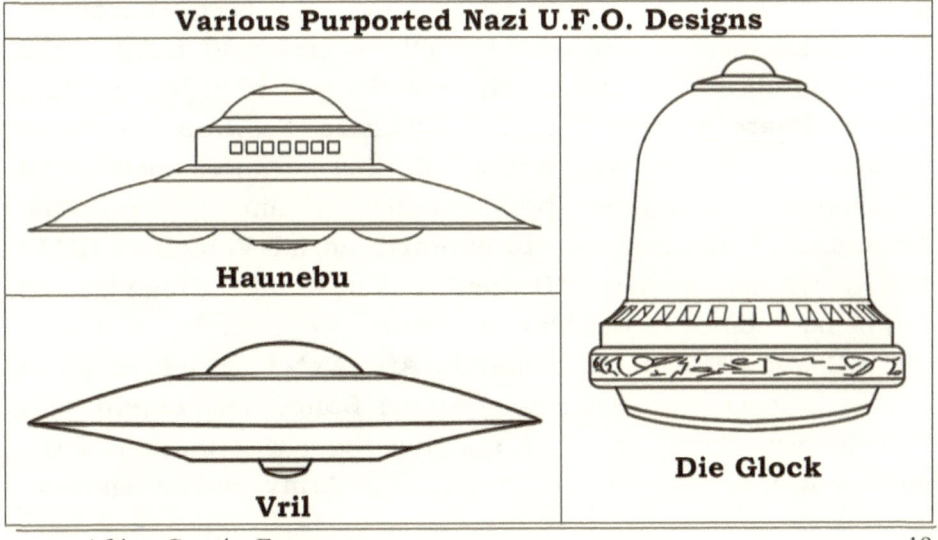

Various Purported Nazi U.F.O. Designs

Haunebu

Vril

Die Glock

The **Nazi U.F.O.** crafts were purportedly seen by American Air Force pilots during **World War II** and called by the name **"Foo Fighters"**. It is rumored that the Nazi U.F.O. crafts were one of the *"wunderwaffe"* (meaning **"Wonder Weapon"** in the German language) that if they had been mass produced would have been able to alter the outcome of World War II. The harsh reality of war brings up another important point in the discussion about liberation and its association to flight or "ascension". Recall that one of the definitions of **"Liberation"** is **"the separation from adverse forces"**, and an **"adverse force"** is **"anything that threatens one's survival and well-being"**. Well physiologically and psychologically, anytime someone is confronted with a threat to their survival and well-being, it triggers within them what is called the **"Fight-or-Flight Response"**. The "Fight-or-Flight Response" is basically a binary decision a person makes when confronted with a threat to their liberation; the person decides they will either confront the threat (Fight), or the person decides they will run away from the threat (Flight). However, considering that in the **Art of War**, the **"high ground"** is considered **the most advantageous position**, and considering that an aircraft may have had the ability to change the balance in World War II, then when we speak of threats to liberation, we must consider a third option in the "Fight-or-Flight" paradigm, which is **"DO BOTH – Fight by Flight"**! Fight-by-Flight as the most advantageous position in war is the reason why aircraft have been weaponized and why military Air Forces are formed; to obtain and maintain the "high ground" in war. In war, the only thing more important than **"the ultimate weapon"** is the "**Ultimate Position**", and the "Ultimate Position" is "from above". Further advances such as the utilization of computer programming in aircraft to create **Autonomous Aircraft**, or **"Unmanned Aerial Vehicles" (UAV)** also known notoriously as **"Drones"**, ensures that the high ground can be maintained in war without putting lives at stake. Consider the 4-Classical Elements of **Earth, Air, Water**, and **Fire**, or in modern science, the 4-Stages of Matter **Solid, Gas, Liquid**, and **Plasma** respectively, then we see that there is a branch of the military to wage war on each front. The **Army** mainly fights on

Earth or Solids, the **Navy & Marines** are for **Water or Liquids**, the **Air Force** is for **Air or Gas**, and they all fight with **Fire (firearms)** or plasma. When a military branch has the ability to fight **in** Fire or Plasma rather than fight <u>with</u> fire (plasma), then this will change the military balance of power in the world. This is part of the Ancient African metaphysical **science of the SPEAR and the SHIELD**.

So we see that not only is "Flight" symbolic of liberation, but Flight is necessary to the preservation and salvation of liberation, and flight is necessary to secure and ensure liberation is maintained. **"Salvation"** is defined as "a deliverance from adverse forces", therefore **"Fight by Flight"** can be seen as a form of **"Salvation"**, that is to say, the security, preservation, and maintenance of liberation. Considering that the Egyptian deity Heru was not only a deity of Flight, but Heru was also a deity of Fight (a warrior deity), then we can see **Heru as a symbol of "Fight by Flight" or "Salvation"**. Heru as a symbol of "Salvation" or "the Savior" would be another association between Heru and the Biblical character of Jesus. However, where Jesus was not a physical fighter and prescribed to the philosophy of **"turning the other cheek"** when attacked, the story of Heru suggests that he believed in the concept of *"Eyes for an eye, and Teeth for a tooth"* in war. **Heru** as a symbol of **"Fight by Flight"** is both **"the Savior and the Liberator"** in one. "Salvation" as the need to fight to ensure that liberation is maintained is the answer to the question *"What Happens after Liberation is obtained? Once the prison doors open and the prisoner is set free, then what?"* The answer is, you ensure that you are never imprisoned again, and you ensure that your liberation is never compromised again. You protect what is valuable to you, and if Freedom and Liberation is valuable to you, you will protect it. The fight to obtain liberation (flight) is often called a **"Revolution"**. Also, aeronautically, "liberation" in the form of flight is obtained by way of the **revolutions** of the **propeller blades** of **helicopters** and **rotor-aircrafts**. The **Revolution of the propeller blades** of helicopters and rotor-aircraft create

whirlwinds which provide rise and uplift. Amongst the Yoruba people of Africa, **OYA** is the name of a **warrior goddess Orisha** of the **whirlwinds (revolutions)** and cyclones. So since we recognize that liberation and salvation is made possible through "**Fight by Flight**", and that flight and uplift can be achieved through the **whirlwinds** created by **revolutions** (of propellers), then this gives new meaning to the **revolutionary** leader **Marcus Garvey's** famous saying "**Look for me in the Whirlwinds.**" Considering that liberation, survival and well-being can be considered a form of "**Heaven**", then "Salvation" or "Fight by Flight" is indeed **a way to "go to Heaven"** or a way to obtain Liberation. **Ascension, and all its many forms and symbolisms, is the way to Heaven**.

This book has been entitled "***Heru-copters: African Aeronautical Ascension***" because it intends to teach and demonstrate the sciences of **Aviation**, **Aerodynamics**, and **Aeronautics** using stories, concepts, and principles associated to the Ancient African deity name **Heru**, as well as other African deities related to the wind and the sky such as the Yoruba Orisha named **Oya**. Affectionately, we can consider **H.E.R.U.-C.O.P.T.E.R.** as an acronym for "**Helicopter Enabling Revolutionary Uplift, Created On Principles To Elevate, Rise, and Soar**". This book is a continuation of the work and dedication of "**African Creation Energy**" to encourage, support, advance, and promote African Creativity, Inventiveness, and Ingenuity for the purpose of developing, engineering, forming, formulating, innovating, inventing, designing, building, and creating any materials, structures, machines, devices, systems, and processes needed for survival and well-being by African people for African people. With that said, the importance of Aeronautics is emphasized in this book.

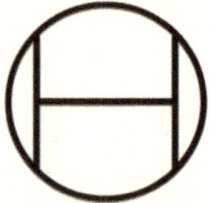

Heru-Copter Launching and Landing Point

2. African Aeronautics

When the "History of Aviation" (or the History of pretty much any modern scientific concept for that matter) is researched in the Western world, often you find discussions from Greek and European mythology are included in the conversation. For example, most articles and textbooks which discuss the **"History of Aviation"** will include the story about **Icarus** from Greek mythology. In the mythological story, Icarus' father Daedalus builds Icarus a set of wings using bird's feathers and bee's wax. Daedalus warns Icarus that when using the wings to fly, Icarus should not fly too close to the sun, or else the bee's wax would melt, and Icarus would fall out of the sky. In the story, Icarus' ego and pride causes him to want to see how high he can fly, so Icarus ignores his father's warning, flies too close to the sun, and the wax on the wings melts, and Icarus falls to his death. The story of Icarus is an obvious mythological allegory about the potential consequences of excessive pride and ambition; however the mythological story is still included in the conversation about the History of Aviation.

Even more surprising, when the history of modern aeronautics is researched, discussions about inventions which did not actually work and fly, but were merely imagined to have worked by their inventor, are included in the conversation about the history of modern aeronautics. Case in point: **the flying machines of Leonardo da Vinci**. Leonardo da Vinci's flying machines include drawings and sketches for a human-powered **hang-glider**, a **parachute**, an **Aerial Screw helicopter**-type of aircraft, and several plans for human-powered **Ornithoper** (flapping-wing) devices which he thought all would be capable of flight. However, there is no evidence that Leonardo ever built any of the flying-machine devices that he drew. And, many years later when people did attempt to build the flying-machines based on Leonardo's designs, none of the devices were capable of flight; they did not even lift off the ground. Leonardo's flying-machines did not fly; they were simply **drawings which were merely representations of the concept** of human-powered flight. However, the flying machines of Leonardo are still mentioned in the conversation about the history of Aeronautics.

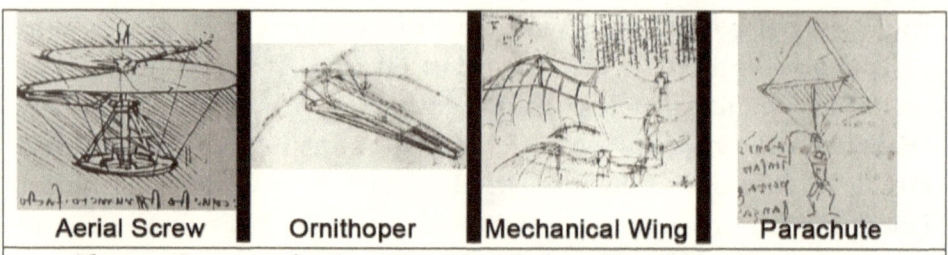

| Aerial Screw | Ornithoper | Mechanical Wing | Parachute |

Above: Leonardo Da Vinci's Flying Machine Drawings

We find conversations about mythology and magic when we study the historical origins of other areas of technological development. For example, when the history of **Robotics** is researched, we find conversations about **clay Golem statues brought** to life by "*magic Hebrew words*" written by Jewish Rabbis. When the history of **Chemistry** is researched, it goes back to the proto-science of **Alchemy** which goes back to mythological figures such as the Greek **Hermes** and the Egyptian **Thoth** (the first Alchemist). If mythology and magic qualifies in the discussion of the historical origin of Robotics, then surely no one should have a problem about the discussion of **African Voodoo Dolls which come to life by way of magic** included in the discussion about the historical origin of Robotics. If a **drawing which is never built** and **merely represents the concept** of flight can be included in the discussion about the history of Aeronautics as is the case with Leonardo da Vinci's flying machines, then surely no one should have a problem if the **"Dendeara Lights"** from Ancient Egypt are included in the conversation about the history of Electrical Lighting because 1) all Light is Electro-magnetic Radiation, 2) the objects drawn in the pictures of the "Dendera Lights" from Ancient Egypt were **actually built**, and 3) the text of the Medu Neter Hieroglyphics associated with the Dendera

Lights explicitly says the objects were used for "Light" (or at the very least, representations of the concept) as is evident in the translation of the text below: "**Re-Har-Sema-Tawy** *is alive with* **gloss** *in the sky (and) lives at the day of the New Year celebration. He* **lights up** *in its house in the night of the child in his nest, by* **donating the light** *to the country from the birth bricks.*" ~Source: François Daumas (French), Wolfgang Waitkus (German), translation of The South Lower Crypt Wall of the Temple of Dendera, Banner across the top

Since the precedent of including mythological, magical, symbolic, and representative concepts in the discussion of the historical origin of scientific principles and technological development has already been established, and in order to be balanced and fair, and not to be biased, or promote a double standard, then it is imperative that the appropriate associated concepts found in African and other non-European cultures be included in the discussion of the historical origin of scientific principles and technological development. But more specifically, if it is a scientific fact that the historical origin of human life started on the African continent, then surely African mythology, African magic, and African symbolism should definitely begin any discussion on the historical origin of scientific principles; and as it relates to this book specifically, **the African origin of Aeronautics**.

The origins of Human inventions for flight start with the study of Nature, so before discussing the history of human aviation and aeronautics, we must first discuss the **Aviation and Aeronautics found in Nature**. When we think of technology, we often think of electronic or mechanical gadgets and gizmos, but broadly speaking, **technology** is the **application of knowledge**, and **knowledge** is the **receipt and storage of information**. Therefore, a plant that turns its leaves towards the sun to receive more sunlight is a form of technology because the plant received information that more sunlight was in a particular direction, then the plant made use of the information that it received by turning its leaves in the direction needed to receive more sunlight. A Lotus flower that opens up in the morning to receive more sunlight and closes at night once the sun has set is an example of the **Technology of Nature**. Plants in Nature have also applied the sciences of Aviation, Aeronautics, and Aerodynamics to create technologies such as the **Samara Seed** which improves the dispersal of the plant's seed across longer distances, and improves the plant's chances of **survival and well-being** through reproduction. The Samara seed is a **winged-seed** which flies through the air with an Aeronautical design similar to that of a **Helicopter's propeller**. Due to the Aeronautical design of the Samara seed, it is able to fly through the air and travel great distances. The Samara seed is found amongst

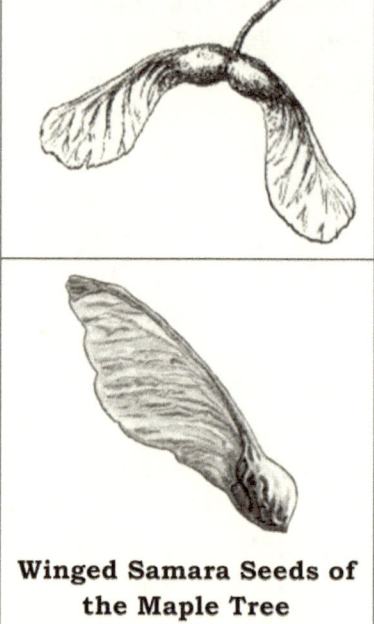

Winged Samara Seeds of the Maple Tree

plants such as the **Maple Tree** and the **Elm Tree**.

When we think of Aeronautics and Aviation in Nature, we immediately think of Birds. Birds are the most obvious and iconic symbol of flight, and have historically been one of the primary motivators behind Human desires to invent flying machines. The ability to fly has only evolved in 4 animals: Insects, Pterosaurs (Winged-Dinosaurs), Birds, and Bats. **Insects** were the first to fly around 350 million years ago, followed by a variety of winged-flying dinosaurs or **Pterosaurs** 200 million years ago. **Birds** began to fly 100 million years ago, and the last of the Avian flying animals are **Bats** which evolved the ability to fly 60 million years ago. Bats are the only mammal which has evolved to have the ability to fly.

These flying animals, called **Avians**, developed the ability to fly independently, which is why there is variation in the wing design of each of these animals. While Human beings studied the flight mechanics of Avian Flying Animals to learn "**How to Fly**", there are still only theories which exist as to "**Why**" these animals evolved the ability to fly, but all of the theories fall under the category of "**survival and well-being**". The theories for why avian flying animals evolved the ability to fly include:

- to become a better predator from the "high ground" or the sky
- to escape predators on the ground
- transportation and the ability to move from place-to-place
- to get access to food that was high up in trees

Evolution of Avian Flight

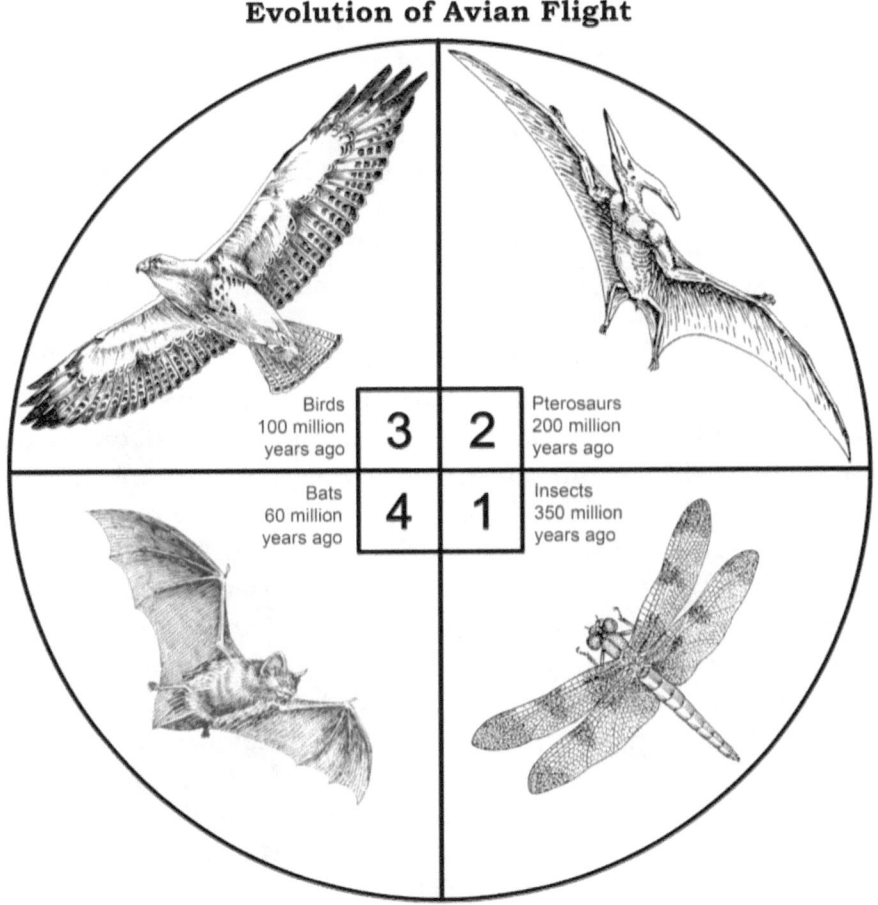

Birds 100 million years ago	**3**	**2** Pterosaurs 200 million years ago
Bats 60 million years ago	**4**	**1** Insects 350 million years ago

In addition to **soaring** and **gliding**, another unique flight characteristic is the ability to **hover** in the air in place. **Hummingbirds** and insects like **Bumblebees** are perhaps the most well-known Avians with hovering abilities, but also some **Bats** are capable of hovering in the air in place by creating **swirling vortices** in the air with their wings. Also, unbeknownst to most people, **Kestrel Falcons**, also known as **Wind-hovers**, have the ability to hover in the air about 45 feet (15 meters) off the ground when hunting and then swoop and dive down to capture their prey. Another significant characteristic which comes with the ability to fly is the ability to travel with phenomenal **Speed**. The fastest flying animal in Nature is the **Peregrine Falcon** who can travel at speeds over 200 miles per hour (322 km/h). Also, Falcons are **tool users** and are one of the **Builders of Nature** in that Falcons build nests. Indeed, perhaps the most remarkable flying creature created and engineered by Nature is the Falcon, because Falcons combine the soaring, gliding, hovering, and speed qualities which make them **masters of the skies**. Rather than waiting for the **Rapture**, you can study Falcons and **Ascend** to the skies right now like a **Raptor** (falcon). Due to these remarkable facts about Falcons, it is not surprising that the Ancient Africans in Egypt first conceptualized the concept of Human flight with metaphors associated with Falcons. The Ancient Egyptian **Aviation god Heru**, who was the son of Asar and Aset, was depicted as an Anthropomorphic Falcon. Heru was also worshipped in the Ancient Egyptian city of Nekhen, which meant "Falcon", and Heru was said to have been able to fly in the sky as it is stated in Pyramid Text Utterance 467. The Ancient Egyptian **Sun God in the Sky** named **RE** was also depicted as an Anthropomorphic Falcon, and Heru and Re were combined into a single Falcon deity of Ascension named **Re-Horakhty** (meaning the Sun god Heru of the 2 Horizons: East and West). The Ancient Africans in Egypt were also the first to conceptualize machines capable of flying through the air. In Ancient Egypt, the **Falcon-Man Sun God RE** was able to fly through the sky on ships called the **Mandjet** (the ship of the morning) and the **Mesktet** (the ship of the evening). Additionally, on the Astronomical Ceiling of the Hypostyle Hall of the Temple of Hathor at Iunet, the Ancient Egyptians personified the Orion and Sirius stars as Human people who were Husband and Wife named **Sahu** and **Sopdet**, with their son named **Sopdu**, riding through the stars on ships. Sahu, Sopdet, and Sopdu are also related to the triad of Asar, Aset, and Heru. The fact that the Ancient Egyptians conceptualized and depicted people riding through the sky on ships and riding through the stars on ships is evidence that the Ancient Egyptians were literally the inventors of the concepts of **Airships**, **Starships**, **Spaceships**, **Aeronauts** (Navigators of the Air), and **Astronauts** (Navigators of the Stars). The etymology of the word "Astronaut" literally breaks down to "star sailor" and the word "Aeronaut" literally breaks down to "air sailor". Not only did the Ancient Africans in Egypt imagine and draw images of Star ships and Air ships, but the Ancient Africans in Egypt also built full-sized versions of these air ships as is evident by the **Khufu Ship** which is now kept in the Giza Solar boat museum. The Khufu Ship was built for the **4th Dynasty Pharaoh Khufu** as a **Solar Barge** or **Solar Barque** to carry the **ascended** king through the **sky** with the sun deity **RE**. The Khufu ship is evidence that the Ancient Africans in Egypt were the originators of the concept of "**heavier-than-air flying ships**". The fact that the Ancient Egyptians depicted people flying through the sky on boats may seem strange, however, when we consider that the movement through, and the movement of, air and water are both covered and considered the same in the modern scientific fields of **Aerodynamics** and **Fluid Mechanics**, then perhaps the Ancient African sky ships in Egypt were based on a profound scientific underpinning.

Egyptian Starship

Egyptian Sky Ship - Solar Barque

The relationship between the aerodynamics of flying through the sky versus swimming in the water can be observed in another artifact from pre-Columbian Incan culture.

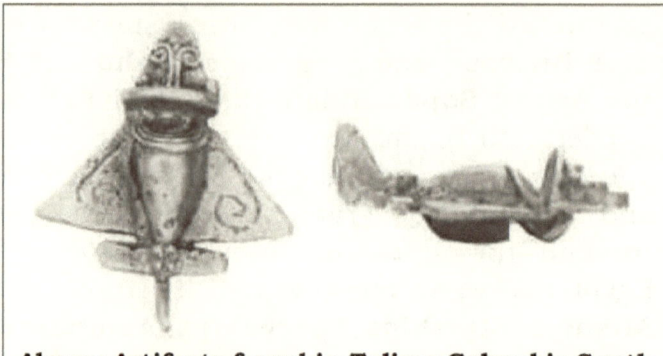

Above: Artifacts found in Tolima Columbia South America - Stylized Fish or Ancient Aircraft?

Referred to as the **"Tolima Fighter Jet"**, because the artifact was found in the city of Tolima in Columbia South America and also because the artifact resembles a modern day fighter jet, the small gold statue is believed by some to be a stylized fish whereas others believe the artifact is an ancient aircraft. In 1997, German Aviation Engineers who were proponents of the "ancient aircraft" theory built a scale-model replica of the artifact to demonstrate that it could indeed fly. Thus, if the artifact is indeed a stylized fish, it demonstrated the aerodynamic relationship between air and water. Also, flying fish are indeed a reality of nature. The scientific relationship between air and water in aerodynamics and fluid mechanics gives new meaning to the idiom "...*in the sky, flying with the fishes, or maybe in the ocean, swimming with the pigeons.*" Another pre-Columbian culture, the **Aztecs**, had what can be considered a type of **"Air Force"**, and elite group of soldiers they called "*cuāuhtli*" meaning **"Eagle Warrior"** which was an exclusive rank in the Aztec Military. Another elite class of the Aztec Military was called the "**Jaguar Warriors**". The combined uniform of the Aztec Jaguar warriors wearing Jaguar print cloth and the Aztec Eagle warriors wearing bird's feathers is identical to the depictions of the **African Orisha Oxossi** who was a deity of war and hunting in the Yoruba religion. **Oxossi** is said to be the master of all air attacks able to deliver swift justice from above; just like an Air Force military campaign.

Modern aircraft fly with vertical tails similar to fish that swim through the water, whereas birds are able to fly through the sky without the use of vertical tails. Another interesting relationship between traveling in air and traveling in water is that Cephalopods like **octopuses** and **squids** propel themselves through water using **Natural Jet propulsion technology** much like modern day aircraft propel themselves through the air using man-made jet propulsion technology. The air ships, star ships, space ships, and solar boats of Ancient Egypt gave rise to Solar Chariots, Sun Wheels, Winged Sun Discs, Flying discs, and Flying saucer concepts in cultures around the world over time. In Ancient times, Pyramids were built on the ground as temples dedicated to a mythology centered around the light of the sun disc, and reflected the positions of various star discs to send the pharaoh's soul to the stars. In modern times, a new yet similar mythology exists where flying discs (flying saucers or U.F.O.s) descend to earth and take people up in a pillar of light to the stars.

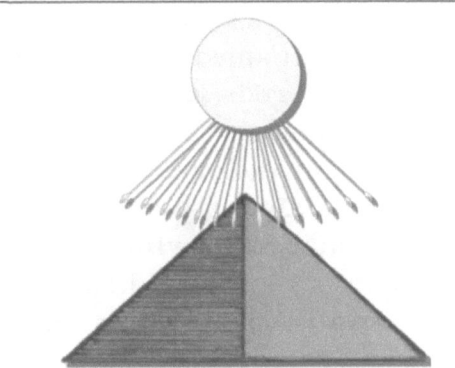	
A disc of Energy (light) above a pyramid of Matter	**A disc of Matter above a pyramid of Energy (light)**
Above: Mythologies related to Discs over Pyramids	

Over time, the concept of flying discs descended from **Heru Behutet**, the winged sun disc of Ancient Egypt and the **Aton** sun dis of Ancient Egypt, to the winged sun disc ridden by the 9th century BC deity named **Ashur** in **Assyria**, to **Faravahar**, the winged disc ridden by **Zoroastrian** deity **Ahura Mazda** in Persia, to the flying discs reported in modern U.F.O. sightings.

This is not to suggest that the winged sun discs in ancient times were U.F.O.s or piloted by extraterrestrials, although the Sun is an extraterrestrial (something outside of earth) by definition. This is to show a similarity in themes in mythology across time, and perhaps the U.F.O. flying discs sighted in modern times were built by humans here on earth who were inspired by the stories of flying discs from ancient times. Considering the hovering and degrees-of-freedom of movement often reported by U.F.O.s is very similar to the hovering and degrees-of-freedom of movement found in human constructed helicopters and multi-copters, and also considering that ancient **sun wheels**, **sun crosses**, and solar symbols resemble both helicopter propellers and flying discs, it is not unreasonable to think that U.F.O. flying discs have a human designed and constructed terrestrial origin.

In addition to Heru, the Ancient Africans in Egypt had other deities with **wings** alluding to the ability of human flight including the deities **Aset**, **Neb-Het** (Nephthys), and **Ma'at**. The Ancient Egyptians also had human deities who lived in the sky including the goddess of the starry sky named **Nut** and the god of Air named **Shu**. Stories of sky gods and humans routinely ascending and descending from the heavens are replete throughout African culture and mythology. Amongst the **Yoruba** people of Nigeria, the **Orisha Oya** is a goddess of the Air who can ascend into the sky on a **Whirlwind**. Also, amongst the **Yoruba** of Nigeria, **Olorun** is the sky god ruler of the Heavens whereas amongst the **Akan** people of Ghana, the sky god is named **Nyame**.

In Mali West Africa, there are stories of the **Tellem** people, **Aero-pygmies** who could fly up to high places and live on the side of cliffs and on top of mountains. A flying Tellem pygmy figure was also included on the flag of Mali up until the year 1961 where it was removed because of Muslim influence in Mali.

**Flying Tellem
Aero-Pygmy Figure
on the 1959 Mali Flag**

The Evolution of the "Flying Disc" Concept Over Time

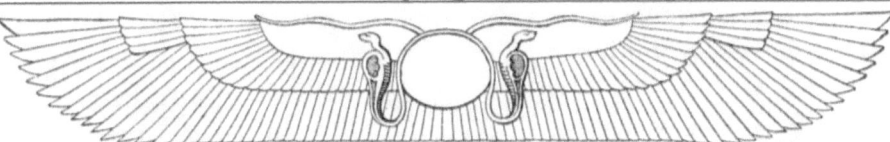

Heru Behutet - Egyptian Winged-Sun Disc (Re-Horakhty)

Ashur – Assyrian Mesopotamian God Riding a Winged-Disc

Faravahar, Ahura Mazda, Zoroastrian God in a Winged Sun-Disc

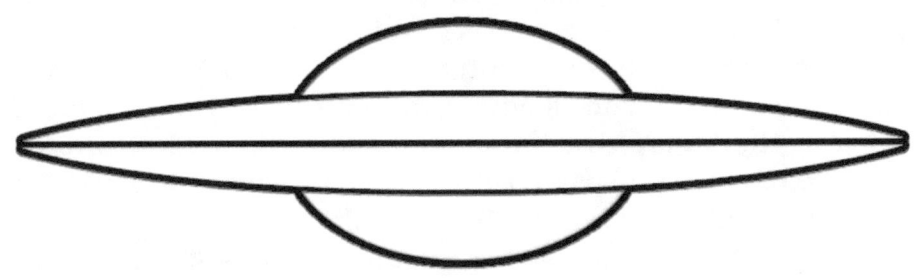

Abstraction of Typical purported U.F.O. "Flying Disc"

The original "flying disc", the Ancient Egyptian Winged Sun Disc called Heru Behutet, was one of the many aspects of the African Aviation deity Heru, along with titles such as "Heru the Elder" and "Heru the Great". As previously mentioned, Heru was also a warrior deity. Heru won the battle to avenge the death of his father, but during the fight, Heru's eye was wounded. The "Eye of Heru" became a revered symbol in Ancient Egypt, and as it relates to modern Aviation, the "Eye of Heru" can be considered analogous to "an all seeing eye in the sky" mounted on surveillance aircraft. **Heru Behutet**, the winged sun disc, is the African origin of the "Flying Disc" or "Flying Saucer" concept. If it is true that the Nazis of Germany did in fact develop flying disc aircraft, then the origin of the concepts and symbols can be traced back to Africa. First, consider that the flight characteristics of purported flying disc U.F.O.s are very similar to that of helicopters in that both aircraft are capable of hovering, and moving in any direction. Also, consider that if the Nazi swastika was animated by spinning, then it would resemble a spinning helicopter propeller and flying disc. The actual scientific name for a spinning helicopter propeller is an **Actuator Disc**. The spinning swastika symbol is also called "**The Hammer of Thor**" with **Thor** being a Norse **Blacksmith** deity. The swastika has its origins in Hindu, Buddhist, and Eastern philosophies, and the swastika is also related to various **sun crosses** and **solar wheel** symbols around the world including the **Dikenga** symbol from the African Congo region and the **Ankh** symbol from Ancient Egypt. The **Ankh** symbol can be found in the tool (hammer) held by the **Ancient African Blacksmith** deity named **PTAH** (sometimes pronounced as "**TAR**"). It is apparent to see how the swastika and the hammer of Thor was taken from the Ankh and the hammer of Ptah (Tar). The way Ptah relates to Aviation, Flight, and the deity Heru is explicitly stated in the **Memphite Theology** on the **Shabaka Stone** on line 53 where it states "...***HERU*** *had taken shape as* ***PTAH***...". The relationship between PTAH and HERU is also found in that in time, PTAH was merged with another **Blacksmith** deity named **Sokar**, who was a Hawk-headed or Falcon-headed deity like Heru. Eventually over time, Ptah-Sokar was merged with

Asar, the father of Heru, to become the Asar-Ptah-Sokar who was depicted as a dwarf (recall the Aero-pygmy flying dwarfs called Tellem in Mali). With Ptah being a craftsmen deity and a symbol of work, and Heru being a symbol of Flight, and Flight being a symbol of heaven, tranquility, and peace, then it is not surprising that in the Ancient Egyptian **Medu Neter** language, the word for "Peace" is HOTEP (HTP) is the reciprocal of the word PTAH (PTH). The phrase "Ptah Hotep (PTH HTP)" or "Hotep Ptah (HTP PTH)" can be considered to represent the concept of "Peace if you're willing to work for it", "freedom if you're willing to struggled for it", "liberation if you're willing to fight for it", *fawohodie ene obre na enam.*

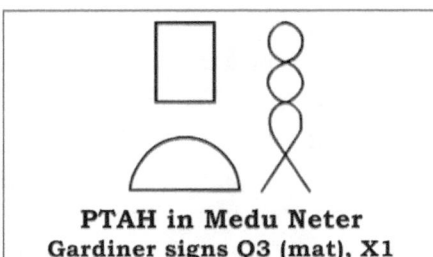

PTAH in Medu Neter
Gardiner signs Q3 (mat), X1
(bread loaf), V28 (twisted rope)

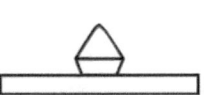

HOTEP in Medu Neter
Gardiner sign R4
(bread loaf on top of mat)

PTAH is further associated with HERU in that the winged Sun Disc *Heru-Behutet* was worn around the neck of the **Apis bull**, an animal said to be sacred to PTAH in Ancient Egypt. The Apis bull of Ptah was kept in a temple in the same city of Memphis in Ancient Egypt as the sacred temple of Ptah called "Hut-Ka-Ptah". It is interesting to note that the Ancient Egyptian word **"Ka-Ptah"** is similar in phonetics to the suffix **"-copter"** which is abbreviated from, and found in the word

Mesopotamian sky god Anu in a winged-disc on a Bull

"Helicopter". There is also evidence that the Ancient Mesopotamian deity **Anu**, god of the **heavens**, was depicted riding on a **Bull** while inside of a **winged-disc**.

Heru, the African Aviator, had 4 sons who along with an association to animals were also associated to the Navigational directions of North, East, West, and South (N.E.W.S.). Spell 148 of the *Per-em-Heru*, the "**Book of Coming forth by Day**" also known as "*the Book of the Dead*", directly associated all four of the sons of Heru with "**the pillars of Shu**" or "**the Pillars of Air**" and also with cardinal directions. The son of Heru named **Duamutuef** was also called a "**rudder of heaven**", also known as a **Vertical Stabilizer** on modern **Airplanes**, and for this reason, the 4 Sons of Heru will be used to represent various principles in the Aerodynamics and Aviation dynamics section of this book. In the Pyramid Texts, **the 4 sons of Heru are described as assisting the king with his Ascension up to Heaven**. The 4 sons of Heru were also used as Canoptic Jars to store body parts of a deceased person after mummification during the soul's ascension into heaven. It is interesting to note that there are 4 Canoptic Sons of Heru and also 4 Synoptic Gospel writers of the life of Jesus (Matthew, Mark, Luke, and John) which is another synchronicity which exists between Heru and Jesus.

THE 4 SONS OF HERU			
Name	**Representative Body Part**	**Representative Animal**	**Representative Direction**
Dua-mutuef	Stomach	Dog	East
I-m-sety	Liver	Man	South
Qebehsenuef	Large Intestines	Hawk	West
Hapi	Lungs	Baboon	North

The African conceptualizations of Human flight in the form of the deity Heru in ancient times, gave way to the participation of people of African descent in the genesis of modern Aviation in the early 1900s on up to the present.

In 1912, **Emory Conrad Malick**, at the Curtiss Aviation School at North Island, San Diego, California, became the first trained African American pilot on record to date.

Eugene Jacques Bullard, from Columbus Georgia U.S.A., travelled to France to escape racial discrimination in America, and became the first black fighter pilot when he joined the 1st Regiment of the Foreign Legion during World War I in 1917.

In 1921, **Bessie Coleman** became the first African American to hold an international pilot license and she was the first licensed female black pilot in the United States

William J. Powell was an Electrical Engineer, soldier, licensed aviator, aeronautical engineer, and **Black Aviation Revolutionary Activist**. In 1928 he was accepted at the Los Angeles School of Flight, and in 1929 he established the "**Bessie Coleman Aero Club**" for **Black Aviators**. In

1931, he organized the first all **Black Air Show** in the United States; which had over 1500 people in attendance. Powell also created a school to teach aeronautical engineers, mechanics, and pilots. Powell was also a writer and author who published the first African American trade journal called the *Craftsmen Aero News*, and in 1934 he published a book entitled "**Black Wings**", which was a fictionalized account of his own life written for the purpose of inspiring African Americans to enter into the field of Aviation as not only pilots but also as designers, engineers and mechanics. He called for people of African descent to "*fill the air with black wings*". In "**Black Wings**", he wrote:

> "*I do not ally myself with the Negro who begs a White man for his job. I ally myself with that young progressive Negro who believes he has the brain, the ability, to carve out his own destiny.*"

The **Tuskegee Airmen** were a group of 996 African-American military pilots who graduated from Tuskegee Army Airfield between 1941 to 1946, and served as the first African-

American military aviators in the United States Army Air Force's 332nd Fighter Group, Squadrons 99, 100, 301 and 302, and the 477th Bombardment Group during World War II.

On February 12, 2009, **Captain Rachelle Jones** and **First Officer Stephanie Brown Grant** made history for being the first all African-American female crew to operate a commercial jet revenue flight as pilots of Atlantic Southeast Airlines (ASA)

flights 5202 (Atlanta to Nashville) and 5106 (Nashville to Atlanta).

Just as the Africans in Ancient Egypt conceptualized being able to fly through the stars on ships in the form of deities like Sahu (Asar) and Sopdu (Heru), this conceptualization has given way to the participation of people of African descent as Astronauts in modern times. Many of the African Astronauts are not only Aviators capable of flying aircraft and spacecraft, but are also scientists and engineers knowledgeable of the design, development, and creation of aircraft and spacecraft.

Arnaldo Tamayo Méndez was the first person of African descent in modern times to travel into space. He traveled into space with the Soviet Union's Cosmonaut Soyuz 38 spaceflight mission in 1980.

Dr. Guion Bluford, an aerospace engineer and Air Force Colonel, became the first African American in space as a member of the crew of the Space Shuttle Challenger on the mission STS-8 in 1983.

Dr. Ronald McNair, a physicist, would have been the 2nd African American in space, but was one of the 7 crew members to die when the Space Shuttle Challenger exploded 73 seconds after liftoff on January 28th, 1986 during its 10th flight mission STS-51-L.

Dr. Mae Jemison, a physicist, was the first African-American woman to travel in space on Space Shuttle Endeavour on September 12, 1992.

Robert Curbeam, a Naval Flight Officer and Aeronautical Engineer, currently holds the record for the most spacewalks during a single spaceflight when he completed four spacewalks during the STS-116 mission. During his career, Curbeam has done a total of 7 spacewalks to date.

On Space Shuttle mission STS-121, **Stephanie Wilson** (above left), an aerospace engineer, became the second African American woman to go into space, and on Space Shuttle Discovery mission STS-116, mission specialist **Joan Elizabeth Higginbotham** (above right), an engineer, became the 3rd African American woman to go into space.

It is good to study history and ancient religions, it is ok to study mythology, it is ok to entertain yourself with science-fiction; however, you do yourself an disservice if you have more knowledge about ancient, mythological, or fictional aircraft, spacecraft, or astronauts than you do about real, modern, and active aircraft, spacecraft, aviators, and astronauts. Mythology and Science-fiction can motivate young people to become scientist and engineers, and many Scientists and Engineers choose to entertain themselves with Sci-fi and fantasy movies and literature. **Dr. Mae Jemison**, the first African American woman to travel into space, was motivated by Science-Fiction movies like **E.T.** (the extraterrestrial) and television show characters like **Uhura** from **Star Trek**, to join **NASA** and become an astronaut. **Dr. Ronald Mallett**, an African-American Theoretical physicist, was motivated to become a physicist by reading the Science-Fiction novel entitled **"The Time Machine"** by H.G. Wells as a child. If you are fascinated by U.F.Os, extraterrestrials, and historical mythologies, then allow your fascination, captivation, and preoccupation with mythology, fiction, and fantasy to be transformed into activation energy to take part in the study, design, development, and creation of aircraft and spacecraft. The invention of the internet has made communication with people all over the world easier than ever, and likewise the invention of aircraft has made traveling to anywhere in the world possible. In the future, air travel to any place in the world will be even more common place, and the world will seem much smaller. It is important that Africans and people of African descent not only be consumers of airline tickets, or users of aircraft in the form of passengers and pilots, but also be creators, designers, and inventors of our own aircraft and spacecraft. Religious and Spiritual beliefs are perhaps one of the strongest motivators of people's actions. Religious and Spiritual leaders such as **Noble Drew Ali**, **Elijah Muhammad**, and **Malachi York** have all introduced concepts related to aircraft and spacecraft to the followers of their respective doctrines, perhaps as a way to motivate their followers to become creators and developers of aircraft and spacecraft. Even the name of the "**Tribe of Shabazz**" which Elijah Muhammad taught was the name of the tribe which African-Americans descended from, alludes to Aviation and Aeronautics in that the name "Shahbazz" means "**Royal Falcon**" coming from a combination of the Persian words "**shāh**" meaning "king" and "**bāz**" meaning "falcon". In 2010, theoretical Physicist

Stephen Hawking made a statement that *"humanity is doomed unless it takes to the stars"* due to the fact that the Earth and Sun will not always be habitable. Therefore, **Aeronautics and Aerospace engineering are two of the sciences necessary for future survival and well-being**. You cannot just wake up tomorrow and start building spacecraft; it takes time, so it's never too early to be motivated to become Aeronautics and Aerospace engineers. The **Dogon** people of **Mali West Africa** believe their Ancestors came from the **Sirius binary solar system**. If you believe your Ancestors came from another solar system and were able to travel to Earth from another solar system, then you should be motivated to become Aeronautics and Aerospace engineers like your ancestors; if they can do it, you can do it. If you believe in the spiritual concepts of **Astral-projection** and **Astral Travel** where it is said that one's spirit is able to be projected to the celestial or astral plane, then you should also be motivated to study Aeronautics and Aerospace engineering which enables one's physical body to actually ascend and be projected to and travel through the heavens and the stellar, celestial, and Astral plane. **Africa is wherever Africans are located in our RIGHT MIND**; this includes on other continents outside of Africa or on other planets outside of Earth. The universe is yours; you can be **Afro-centric** and also be **Astro-centric**. The day will come when the Earth will become uninhabitable. Rather than being idle and doing nothing waiting for a ship to save you, or relying on someone or something else to save you, allow your historical, spiritual, and religious fascinations to provide you with the initiative to do something productive and constructive for yourself by studying Science, Math, Technology, Aeronautics, and Aerospace Engineering so you can build a ship, aircraft, or space craft, to save yourself and others. When you are able to save yourself and others, you will have become a **Savior**, also known as Jesus or **Heru** in various religious ideologies. However, before you can enter space, you first have to travel through Earth's atmosphere, so Ascension starts with **Aeronautics**.

"Did you ever know that you were a Heru? You're everything you wished you could be. You can fly higher than an Eagle, by controlling the wind beneath your wings"

Question: What did Noble Drew Ali have to say about aircraft and spacecraft?

Answer: Noble Drew Ali spoke about aircraft and spacecraft, which he referred to as an **"apparatus"** capable of taking people up into the air and into outer space, on at least two occasions from the following quotes:

Sister M. Payton-Bey of Temple 4 and 25 said that the Holy Prophet Noble Drew Ali said, *"I have got airplanes, zeppelins, and apparatus. I am going to take my good Moors up in an apparatus on an incline until it's all over with."*

Sister M. Tiggs El of Temple 9 said that the Holy Prophet said that *"at the End of time, those that will be in the apparatus will be able to look down on earth, and see people that you know, fleeing for their life."*

The "Circle 7" - Ascension Apparatus

Question: What did the Honorable Elijah Muhammad have to say about aircraft and spacecraft?

Answer: The Honorable Elijah Muhammad had this to say:

*"The vision of **Ezekiel's wheel in a wheel in the sky is true** if carefully understood...The likes of this **wheel-like plane** was never seen before. You cannot build one like it and get the same results. Your brains are limited. If you would make one to look like it, you could not get it up off the earth into outer space. The similar **Ezekiel's wheel is a masterpiece of mechanics**...His vision of the wheel* included hints on the Great Wisdom of Almighty God (Allah); that really He is the Maker of the universe, and reveals just where and how **the decisive battle would take place (in the sky)**...The present **wheel-shaped plane known as the Mother of Planes**, is one-half mile of a half mile and is **the largest mechanical man-made object in the sky**. It is a small human planet made for the purpose of destroying the present world of the enemies of Allah. The cost to build such a plane is staggering! **The finest brains were used to build it**. **It is capable of staying in outer space six to twelve months at a time without coming into the earth's gravity. It carried fifteen hundred bombing planes with most deadliest explosives** -- the type used in bringing up mountains on the earth. The very same method is to be used in the destruction of this world. The bombs are equipped with motors and the toughest of steel was used in making them. This steel drills and takes the bombs into the earth at a depth of one mile and is timed not to explode until it reaches one mile into the earth. This explosion produces a mountain one mile high; not one bomb will fall into water. They will all fall on cities. As Ezekiel saw and heard in his vision of it (Chapter 10:2) the plane is terrible. It is seen but do not think of trying to attack it. That would be suicide! **The small circular-made planes called flying saucers, which are so much talked of being seen, could be from this Mother Plane**. This is only one of the things in store for the white man's evil world. Believe it or believe it not! **<u>This is to warn you and me to fly to our own God and people</u>**."

~ **Message to the Blackman in America
Chapter 125, Battle in the Sky is Near**

Question: What did Dr. Malachi York have to say about aircraft and spacecraft?

Answer: Dr. York produced many forms of media including books, audio lectures, and video lectures discussing the topic of aircraft, spacecraft, U.F.O.s, and extraterrestrials.

In his book entitled "**Man From Planet Rizq**" Dr. York discusses extraterrestrial beings he calls Rizqiyians which travelled through space from another galaxy to the planet Earth and were the progenitors of Human beings:

"*The Rizqiyian*
Origin: *The 8th Planet, Rizq, in the Tri-Star Galaxy Illywun...*
Appearance: *Overall appearance of a Human Being...*
Nature: *Kind and Loving, Yet Stern.* ***They Claim To Be The Parents Of The Nubians On The Planet Earth****...*"
                    ~ **Man From Planet Rizq, page 111**

3. Aerodynamics:

In modern science, the movement and flow of Gases and Liquids is studied in the areas of **Aerodynamics** and **Hydrodynamics** respectively, which are both sub-fields of a larger area of study called "**Fluid Dynamics**". In modern science, the term "*Fluid*" does not just refer to Liquids, but it also refers to states of matter that can "flow" like liquids, which includes Gases like Air (i.e. "air flow"). The Ancient Africans in Egypt demonstrated their comprehension of Fluid Dynamics, the relationship between Gases and Liquids, in at least two instances: 1) the Ancient Egyptian depictions, mythologies, and ritualistic ceremonies of boats designed to sail on water being said to sail through the air, and 2) the personification of the concepts of Air and Water as a pair of deities who were husband and wife named *Shu* and *Tefnut* in the Ancient Egyptian mythologies. In the Ancient Egyptian mythology of the Ennead, *Shu* and *Tefnut*, who represent Fluid Dynamics and its sub-discipline Aerodynamics, are the Great Grandparents of **Heru** or "**Aviation**".

African Deities of Fluid Dynamics (Aerodynamics), Great-Grand Parents of Heru (Aviation)	
Shu – Air	Tefnut - Liquid

In Ancient Egypt, the personification of the wind and air was the deity named **Shu**. Shu was the husband of the goddess **Tefnut** who represented Liquids, thus, their marriage can be viewed as a representation of "*Fluids*" in general. *Tefnut* was depicted as a woman wearing the feline lioness mask. The relationship and personification of the wind and water can also be found in the West African **Orisha** tradition with the deity **Oya Iansan** representing "winds" and **Yemaya** representing "water".

For the fields of **Aerodynamics** and **Aeronautics**, which are dependent on the presence of **air**, "**the Sky is the Limit**", because once you leave the "sky" and leave the Earth's atmosphere, there is no more air in the vacuum of space, and you have reached the limit of usability of the science. Once you leave the Earth's atmosphere, the science required to travel through space changes from **Aerodynamics** to **Astro-dynamics** and from **Aeronautics** to **Astronautics**. However, there are still some fundamental scientific principles taught in the field of Aerodynamics which are still applicable in space travel which is why the topics of Aeronautics and Astronautics are both covered in **Aerospace Engineering** programs.

Whereas **Ascension** represents the mystical and esoteric wing of the topic of flight, **Aerodynamics** represents the operative and exoteric wing of the topic of flight. Mystical and mythological stories related to ascension give rise to operative and practical aerodynamic inventions. For example, the word **Aeronautics**, which is the science and art of creating crafts capable of flight, comes from the Greek words "*aero-*" meaning "air" and "*-nautics*" meaning "sailing or navigation", which combined literally meant "sailing in the air" – a concept first depicted in the mythologies of Ancient Egypt. There are also other symbols and mythologies from Ancient Egypt that can be used to represent and teach some of the fundamental scientific principles in the field of Aerodynamics.

In the field of Aerodynamics, there are **4 fundamental forces related to flight**: 1) Lift, 2) Drag, 3) Thrust, and 4) Weight.

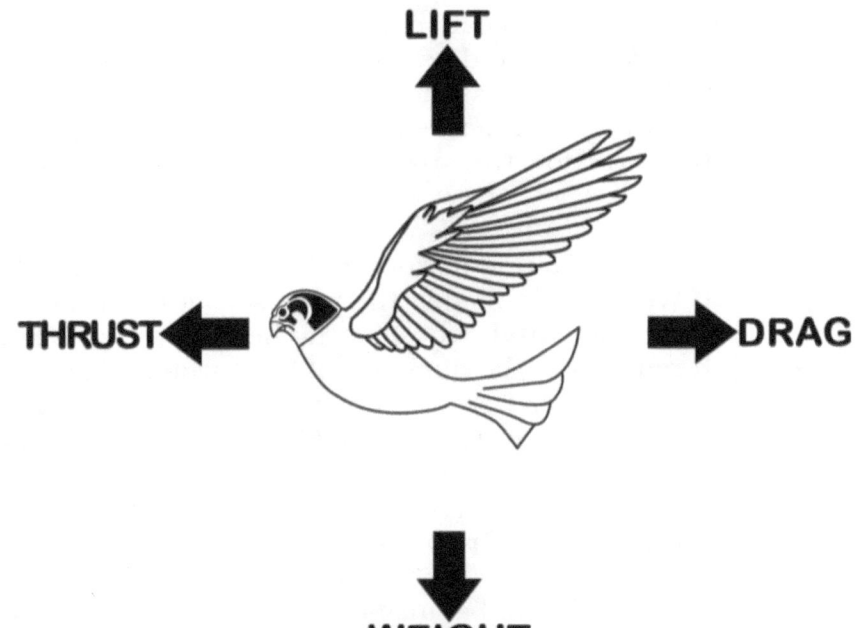

Above: The 4 Forces (Sons) with Heru while in Flight (Ascending)

As controlling these four forces is essential to flight, and Heru is our deity and symbol related to Aviation and flight, then we can use the **4 sons of Heru** as representations of the **4 forces related to flight**.

THE 4 FUNDAMENTAL FORCES RELATED TO FLIGHT (The 4 Sons of Heru)			
FORCE	**Son of Heru Name**	**Animal Symbol**	**Representative Direction**
Lift	I-m-sety	Man	South
Thrust	Qebehsenuef	Hawk	West
Drag	Dua-mutuef	Dog	East
Weight	Hapi	Baboon	North

In our Aerodynamic model to comprehend the 4 fundamental forces related to flight, imagine that Heru is flying towards the West. For our model, the reason why we are orienting Heru to fly towards the West is because the sun travels from East towards the West during the day (rises in the East and Sets in the West), and Heru is also associated with the sun, so there is synchronicity in this model between the direction Heru is flying towards and the direction the sun is traveling towards during the day.

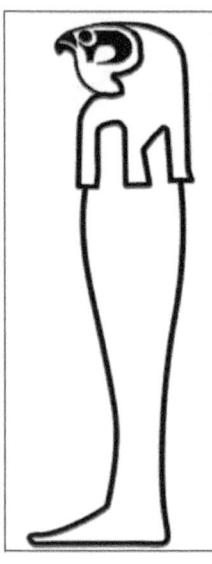

Given the orientation of Heru flying towards the West in our model, then the son of Heru named **Qebehsenuef**, who has the head of a Hawk and represents the direction of West would represent the **Thrust Force**. Simply put, the Thrust Force is the Force which *PROPELS YOU FORWARD*. Scientifically, Thrust is caused by the Law of Nature which states that "**Every Action Has An Equal And Opposite Reaction**" (i.e. **Ma'at** or "**balance**"). Thrust is the Force which occurs as a Reaction to balance out the Force created from the action of mass or air accelerating from an object in the opposite direction. Thrust is not exclusively an Aerodynamic force as there are a variety of methods which can be used to propel an object forward.

The etymological meaning of the word "**South**" is "**sun-side**", and since the Sun is UP in the sky, then the son of Heru who we will use to represent the **Lift Force** in our Aerodynamic model is **I-m-sety** (who represents the direction of South and has the head of a man). The Lift Force is the Force which *RAISES YOU UP*. The Lift Force is Levitation and its opposite is Gravitation. The Lift Force is Levity, and its opposite is Gravity. When fluid in the form of air is flowing past an object, Lift is the upward force perpendicular (at a 90 degree angle) to the oncoming flow direction.

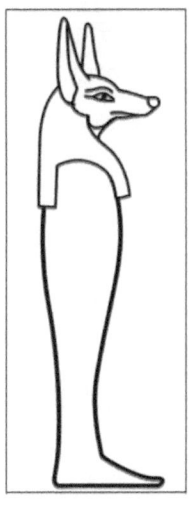

In our Aerodynamic model, since Heru is flying and thrusting towards the West, then the **Drag Force** is the force in the opposite direction, and in the case of our model, towards the East. Since the son of Heru named **Dua-mutuef**, who has the head of a dog, also represents the direction of East, then we will use him to represent the Drag Force. Scientifically, the Drag Force is also known as **Air Resistance**, and is the force acting to oppose the direction of motion of an object. Simply put, the Drag Force is the Force which *HOLDS YOU BACK*.

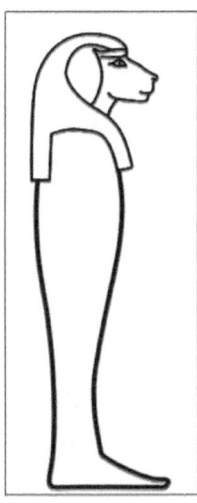

Lastly, the etymological meaning of the word "**North**" comes from words meaning "below" and "from beneath". Since the son of Heru named **Hapi**, who has the head of a baboon, represents the North direction, and the word "North" is related to "below and beneath", then we will use him to represent the last of the 4 forces which comes from "below and beneath" and that Force is **Gravity**. Scientifically, the force on an object due to Gravity is called **Weight**. Simply put, Weight and the Gravity Force is that which *HOLDS YOU DOWN*.

Achieving and sustaining flight requires mastery of combining, controlling, and manipulating the 4 forces of Lift, Thrust, Drag, and Weight. Some of the important combinations of these forces include:

- **Lift-to-Drag Ratio (L/D)**: the amount of Lift produced by an aircraft divided by the drag it creates flying through the air. Higher Lift-to-Drag ratios are most desirable.
- **Thrust-to-Weight Ratio**: the amount of Thrust produced by an aircraft versus the weight of the aircraft. Higher Thrust-to-Weight ratios are most desirable.

- **Lift Coefficient**: the relationship between the lift generated by an aircraft to the density of the air surrounding the aircraft, the velocity of the aircraft, and the area of the aircraft's wing.

The Lift Coefficient is expressed mathematically as the **Lift Equation**:

$$C_L = \frac{2L}{(A \times \rho \times V^2)}$$

where C_L is the Lift Coefficient, L is the Lifting force, ρ is the density of the air, V is the velocity, and A is the area of the wing.

In Aerodynamics, the Wing of an aircraft is called an **airfoil**. An airfoil is a solid object shaped in such a way that when placed in moving air, or moved through air, it overcomes the Drag Force and generates a Lift Force greater enabling the aircraft to fly. Not only are the wings of airplanes airfoils, but even Helicopter propellers are considered airfoils. In fluid dynamics, both the wing of a bird and the fin of a fish are "foils", but the bird's wing is considered an air foil because the bird flies through air, and the fish's fin is considered a hydrofoil because the fish travels through water. In the study of Fluid Dynamics, you can say *Fish "Fly" through the water just like Birds "Swim" through the air*. Imagine you are looking from the end of a wing; there are 4 types of airfoil cross-sections which produce different flight characteristics. The 4 types of airfoil cross-sections of a wing are 1) Symmetrical, 2) Semi-Symmetrical, 3) Non-Symmetrical or Flat-Bottomed, and 4) Under-cambered. The airfoil wings are placed at a certain **Angle of Attack** while passing through the air to provide lift to the aircraft. In Aerodynamics, The Angle of Attack (AOA) is the angle between the airfoil and the oncoming moving air flow. Depending on the airfoil and application, Airfoils should be raised to a 0 to **7** degree Angle of Attack. As a mnemonic device, the **Angle of Attack** can be likened to the **"Angel of Attack"**, the warring **Angel Michael** of religion, who is often depicted flying and "attacking" evil doers during the **War in Heaven**.

4 TYPES OF AIRFOILS		
SHAPE	**NAME**	**CHARACTERISTICS**
	Symmetrical	Same shape on both sides, gives the most speed, good for aerobic flying both right-side-up and up-side down, low drag but gives the least lift
	Semi-Symmetrical	Shaped with higher arch on top than on the bottom, good compromise between lift and aerobic ability, good lift-to-drag ratio
	Non-Symmetrical or Flat Bottomed	Shaped with an arched top and flat bottom, provides good lift at slow speeds
	Under-cambered	Shaped such that the bottom surface is curved up like the top surface, provides the most lift but also has the most drag

When the airfoil is the wing of an airplane, the entire plane must travel forward on the ground at a certain speed, and continue traveling forward at that speed in the air, in order to get the air flowing around the airfoil to generate lift. However, if the aircraft were to remain stationary and the airfoil was made to spin quickly in the air, this would generate lift also, and this is the concept which led to the invention of **Helicopters**. This is why helicopters can take-off vertically while airplanes require runways to build up enough speed to takeoff.

Propellers and Helicopter rotors are airfoils also. In Aerodynamics and Fluid Mechanics, the mathematical model describing how Helicopter propellers provide lift to the aircraft is called **Actuator Disk Theory**. When Helicopter propellers spin, a **Vortex** flow of air is created. Actuator Disk Theory describes the pressure and velocity of the air through the propellers, and provides a mathematical relationship between power (P), thrust (T), air density (ρ), and area of the propeller disc (A), such that:

$$P = \sqrt{\frac{T^3}{2\rho A}}$$

The Yoruba Orisha **Oya** is an African warrior goddess of the wind, hurricanes, tornadoes, and vortexes. If we imagine that Oya's arms are two propellers, and she is wearing a bell-shaped dress, then as Oya spins to create a vortex, the resulting image abstraction is the same as the model used to describe Actuator Disk Theory.

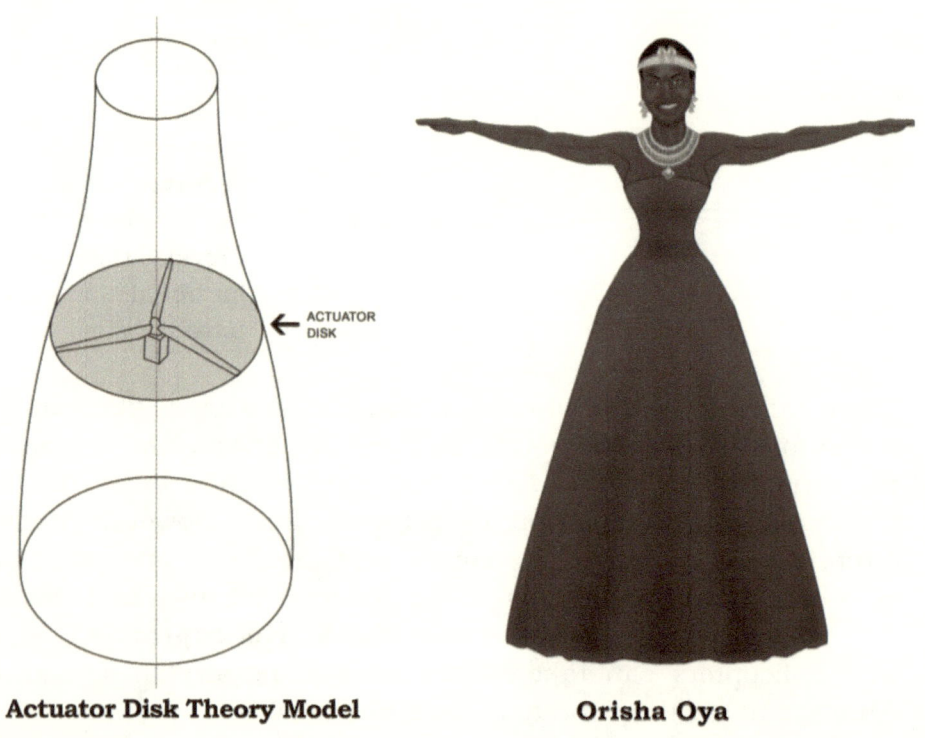

Actuator Disk Theory Model   **Orisha Oya**

When the main rotor of a Helicopter spins, it creates a **torque effect** on the aircraft in the opposite direction. This is because of the law of Nature which states, *"for every action, there is an equal and opposite reaction"*. The tail rotor which appears on most Helicopters provides a Thrust Force to offset the torque effect and keep the helicopter body straight instead of spinning.

Another solution to offsetting the torque effect of the main rotor of a Helicopter is to design the aircraft such that the body, or some other component, of the aircraft spins to offset the torque effect. In Helicopter aircraft with this design, there would be no need for the tail rotor, and the body of the craft could be designed circular, or "saucer shaped", to spin and offset the torque of the main rotor. Aircraft built based on this design also use the **Coandă effect** to generate aerodynamic lift for the craft. The tendency of air, or fluid, to closely flow along a solid surface, and bend and be attracted to the solid surface when the curvature of the surface is subtle and not too sharp is called the "Coandă effect". The Coandă effect is named after an aerodynamicist who obtained a patent in 1934 for a *"Method and **apparatus** for deviation of fluid into another fluid."*

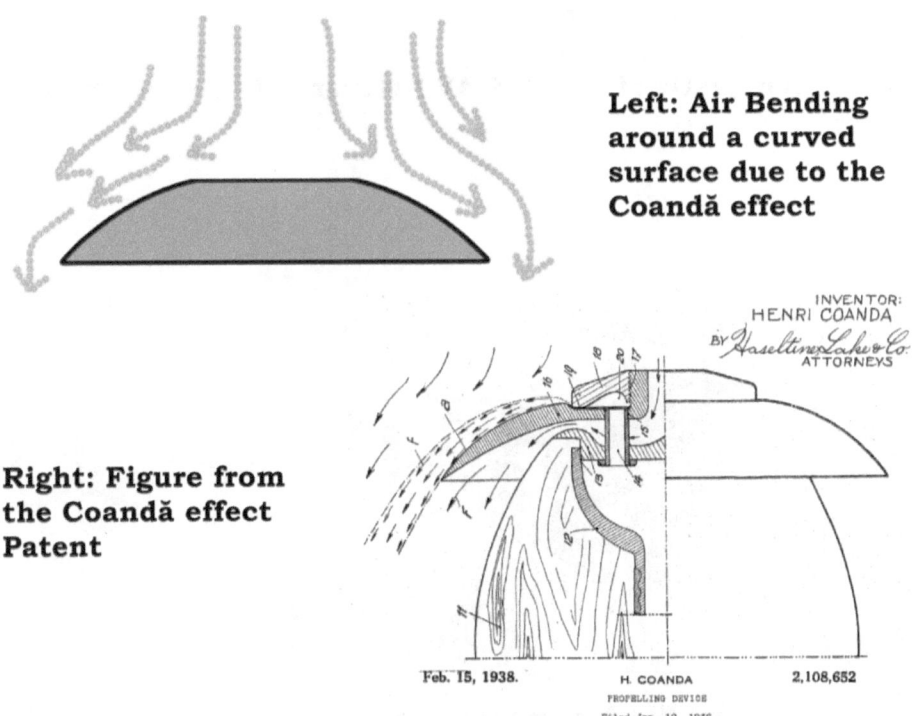

Left: Air Bending around a curved surface due to the Coandă effect

Right: Figure from the Coandă effect Patent

The fact that the Coandă effect causes air to bend down and provide aerodynamic lift was one of the reasons it was one of the principles used to build a real, working **"U.F.O. flying saucer"** here on Earth. In the late 1950s, a Canadian company called **Avro Canada** built a working U.FO. flying saucer called the **VZ-9 AV Avrocar**. In 1956, Avro Canada was commissioned by the United States Military to build a larger supersonic U.FO. flying saucer under the codename **Project 1794**. In 2012, the National Achieves de-classified the schematics of Project 1794, and the details are now available under the Freedom of Information Act.

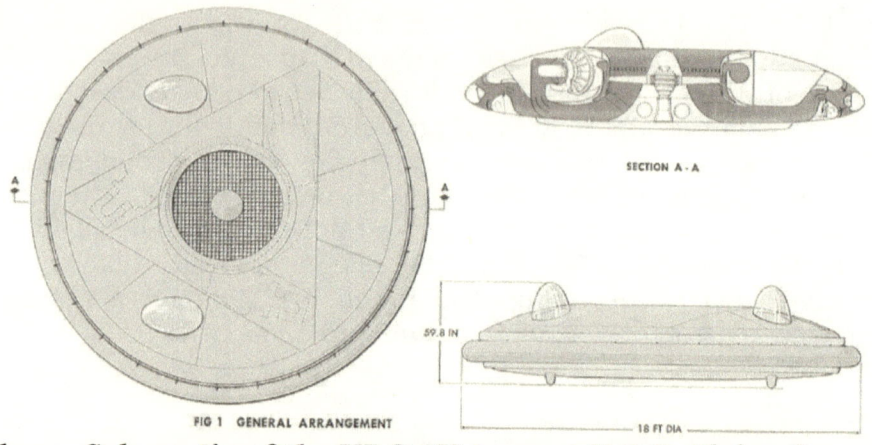

Above: Schematic of the VZ-9 AV Avrocar U.F.O. Flying Saucer

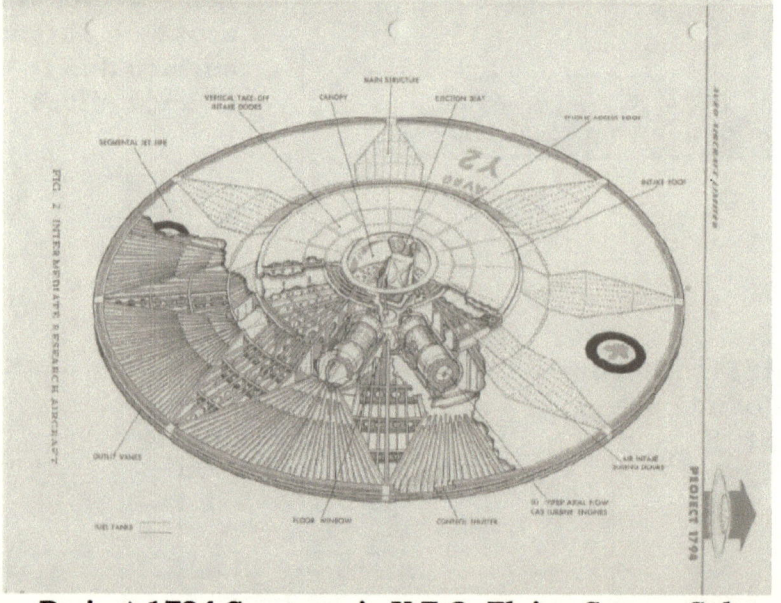

Above: Project 1794 Supersonic U.F.O. Flying Saucer Schematic

The reality is that there is a long and storied history of the development of human-built flying disc and flying saucer aircraft with evidence that overwhelms any claims that these U.F.O. craft are of extraterrestrial origin. Below is a list of some of these flying disc aircraft.

FLYING DISC AIRCRAFT	INVENTOR	YEAR
The Umbrella Plane	William Romme, Harold McCormick	1910
Circular Angular-winged biplane	John Kitchens	1911
Lee-Richards No 1 Monoplane	Cedricc Lee, George Richards	1911
Webb Flying Machine	John Webb	1918
Aerodina Lenticulara	Henri Coanda	1930
Trenn Ducted Fan Aircraft	Paul Trenn	1933
Heinkel-BMW Flying Disc	Ernst Heinkel, Rudolph Schriever	1941
Prague Flying Disc	Giuseppe Belluzzo, Nazi Germany Engineers	1942
Discopter	Alexander Weygers	1943
Omega Diskus	Andreas Epp	1943
Fluid Controlled and Propelled VTOL aircraft	Charles Neumann, Hugo Baca	1945
de Beaumont rotor-wheel flying disc	Francesco de Beaumont	1950
Gyrocopter Flying Saucer	Michel Henri Marie Joseph Wibault	1950
Schnellflugzeug	Heinrich Focke, Wulf	1950
Goodyear Convertible Aircraft	Robert Ross, Goodyear Company	1953
High Performance Ramjet Flying Saucer	Nathan Price, Lockheed Special Projects, USAFB	1953
Triangular Gyrocopter	Michel Henri Marie Joseph Wibault	1953
Dusenscheibe	Heinrich Fleissner	1954
Lenticular-shaped Highly Manoeuvrable flying disc	John C. Fischer Jr.	1954
Project Y	John Frost, Avro-Canada	1954
Streib VTOL flying disc aircraft	Homer Streib	1954
Couzinet Aerodyne RC.360	Rene Alexandre Authur Couzinet	1955
Davis Circular-winged Flying Saucer	John W. Davis	1955
Project MX 1794 Turbo-ramjet flying disc	John Frost, Avro-Canada, USAFB	1955
Project Y2, 9961 Silver Bug	John Frost, Avro-Canada	1955
Doak VTOL ducted fan Aircraft	Edmond Doak	1956
Pa.49	Nicolas Roland Payen	1956
Avrocar	John Frost, Avro-Canada	1957
Weapons System WS-606	John Frost, Avro-Canada, USAFB	1958
XM-4 VTOL aircraft	Paul Moller	1973
Radial Flow Gas Turbine (RFTG) Flying Saucer	Norman Heuvel	1976
M200X	Paul Moller	1989
Geobat	Jack Morris Jones	1990
Rieken Aircraft	William Dee Rieken	2003

Considering that both **flying saucer U.F.O.** craft and **airplanes** are both built and designed right here on Earth, but the flying saucer craft which are built and designed with a **circular shape (O)** are considered **Unknown**, **Unidentified**, **Mysterious**, and **Secret**, but the airplanes which are built and designed with a **cross shape (+)** are **Known**, **Identifiable**, and **Recognizable**, then it may be that the designers of both of these types of aircraft are knowledgeable of the metaphysical association of the **circle being symbolic of "the unknown"**, and **the square or cross being symbolic of "the known"**.

The Circle and the Square (Cross) The Unknown and the Known

Circle Shaped Flying Aircraft are said to be "Unidentified" - Unknown	"Cross" Shaped Flying Aircraft are said to be "Identified" - Known

Regardless if it is a spinning airfoil (as is the case with Helicopter propellers, circular aircraft, and Actuator Disk Theory) or the Airfoil of a fixed-wing airplane, the air passing over and under an airfoil generates the **Lift Force** by creating a pressure differential above versus below the airfoil. Since the top of an airfoil is curved, air traveling over the top of the airfoil has a longer distance to travel than air traveling under the bottom of the airfoil. As the speed of the airfoil (or aircraft) increases, the air moving across the top of the airfoil is moving faster than the air moving under the bottom of the airfoil. The faster moving air over the top of the airfoil is also at a lower pressure than the slower moving air underneath the airfoil.

This pressure differential creates **lift** from the higher pressure (bottom of airfoil) to the lower pressure (top of airfoil). The name given by modern scientists to the explanation of the **Lift force** being generated from the **potential difference** of pressure and energy which exists due to **air flowing** above and below an airfoil is called **Bernoulli's Principle**. The 18th Century Swiss Mathematician and Scientist named **Daniel Bernoulli** is said to have described the phenomenon in his book entitled **Hydrodynamica**. In 1828, during his research on the **strength of the human heart pump, and the flow of blood** through the veins and narrow tubes, Jean Léonard Marie Poiseuille developed an equation called **Poiseuille's Law** by modern Scientists which establishes a relationship between Power, Fluid flow (liquid or gas), and pressure for specialized cases when the liquid or gas fluid is flowing through a pipe or narrow tube.

Bernoulli

Poiseuille

The "**Bernoulli Principle**" which describes a relationship between Force (Lift), Potential Difference due to Pressure, and Fluid Flow, is similar to **Poiseuille's Law** which is a mathematical equation relating Power, Pressure, and Fluid Flow. Poiseuille's Law is part of a set of interdisciplinary analogies between Electrical, Thermal, Hydraulic (Fluid Mechanics, and Aerodynamics), and Mechanical disciplines used by Science and Engineering professors as a form of **Transformative learning and Constructivist Pedagogy**.

The book entitled "**P.T.A.H. Technology: Engineering Applications of African Sciences**" by **African Creation Energy** shows how the ancient African Symbols of the Djed Pillar (meaning stability), the Ankh (meaning life), and the Waas (meaning Power), which appear in the staff held by the Ancient African deity of Engineering named **PTAH**, can be used as analogies to the Mathematical Equation of **Ohm's Law** used in the field of **Electrical Engineering**. The book entitled "**9 E.T.H.E.R. R.E. Engineering**" by **African Creation Energy** shows how the topics of **Thermodynamics**, and **Fluid Mechanics** (**Hydrodynamics** and **Aerodynamics**) can be extended from the discussion of the movement of electrons, using African symbology. When electrons flow through a wire, it causes Heat; a Thermodynamic phenomenon. Heat in turn causes temperature change, which

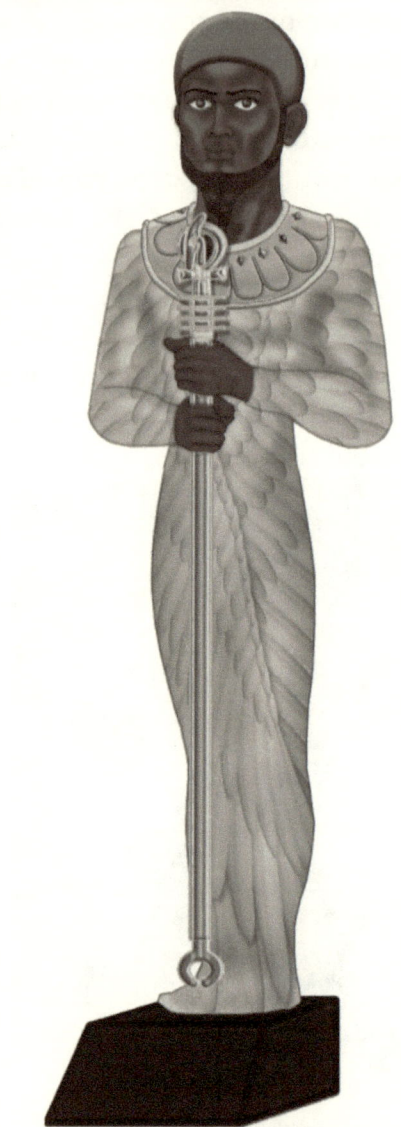

Above: African Creation Deity PTAH shrouded in wings

causes fluid in the form of liquid or gas to flow due to the change of pressure given by the Ideal Gas Law $PV=nRT$ (where

P is "pressure", T is "Temperature", V is Volume, n is the amount or quantity of Gas, and the constant $R=8.314$ J/K*mol). Thus, there is a direct relationship between the movement of electrons, or electricity, to the flow of heat, or Thermodynamics, to the flow of Liquids and Gases, or Fluid Dynamics (Aerodynamics and Hydrodynamics).

In African symbology, the relationship between Electronics and Thermodynamics is depicted in that **PTAH** who is a primary creator (fundamental particle, i.e. **electron**) and holds the Djed, Ankh, and Waas symbols, is married to **Sekhmet** (whose name means "the powerful one") who is associated with the **"power of heat"** or **Thermodynamics**. Sekhmet is the daughter of the Sun deity RE, which makes her a sibling to *Shu* and *Tefnut*, who represent **Fluid Dynamics** (*Aerodynamics* and *Hydrodynamics*). The Ancient Egyptian story of **Shu** (Gases) and **Tefnut** (Liquids) being husband and wife and both being children of **RE** (Sun, Fire, or Heat) establishes a parable that shows relationships between

Above: African Creation Deity
SEKHMET, wife of PTAH

modern scientific concepts like **steam**, **condensation**, and **evaporation** which are transitions between **gases** and **liquids** caused by **heat**. The application of the science of Thermodynamics as it applies to transitions between liquids and gases led to the development of one of the earliest forms of engines called the **Steam Engine**. Modern Science credits the invention of the steam engine to a Greek man named **Hero of Alexandria**. Hero of Alexandria was a member of an Ancient Greek organization called the **Atomist**. It has been suggested that the origin of the name of the Atomist group came from the Ancient Egyptian deity **Atum**. It could very well be that Hero of Alexandria was inspired by Ancient Egyptian science in the development of his Aeolipile. Even the name "**Hero**" is phonetically similar to "**Heru**" of ancient Egypt.

The Ancient Egyptian wind deity Shu is sometimes depicted seated facing the **Ankh**. The Ankh is an Ancient African symbol called "The Key to Life", or "continuous life". The Ankh embodies the unification of dual principles and represents the flow of life giving fluid (called Semen) from the Male Principle at the bottom to the Female principle at the top which is "**the key to human life**" in the

Shu facing The Ankh

Ankh's Biological symbolism, and the flow of Fluids (Gases and Liquids) due to temperature and pressure differences which is the "**key to fluid dynamics**" in the Ankh's Physics symbolism. Hero of Alexandria wrote about the Hydrodynamic principles of the **Siphon** in a book entitled "**Pneumatica**". However, in the book entitled "**A History of Mechanical Inventions**" the author indicates that there is evidence that the Ancient Egyptians during the **18th Dynasty** used Siphons to draw oil out of **canopic jars**, which were usually fashioned in the form of the **Four Sons of Heru**, as early as 1500 BC. Siphons are basically **pipes** or **tubes** that make use of Hydrodynamic principles to cause liquids to flow between two or more containers at different heights. The liquid flows through the Siphon because the liquids in the containers are at different heights and thus have a difference in gravitational pull, **potential energy**, and pressure. Pressure differences in

Hydrodynamic processes cause the movement of liquid in a siphon, and similarly **Blood Pressure** causes the **circulation of blood** through your **veins** which enables **life**. In his article entitled "*Reinterpretations of the Ankh Symbol Part 2*" by **Asar Imhotep**, he discusses the relationship of "bio-mimicry" between the appearance and meaning of the Djed, Ankh, and Waas symbols, to the Spinal Column, Thorax Bones, and Brain stem respectively, and how this relates to the **Cardiovascular processes** which sustain life in the Human body. Considering that the Cardiovascular process in the Human body is a form of **Hydraulics** and **Fluid Mechanics** with the flow or movement of **liquid** in the form of **Blood**, and the flow or movement of **air** in the form of **breath** or respiration, then we can see how the symbols of the Djed, Ankh, and Waas can serve as symbols in Mathematic equations related to Electricity, Thermodynamics, Fluid Dynamics (Aerodynamics and Hydrodynamics), and Mechanical Engineering.

Electrical, Thermal, Hydraulic, and Mechanical interdisciplinary analogies have been developed over time, and have been routinely included as part of curriculums taught in Science and Engineering programs at Universities across the world. The relationship between Power, Current, and Voltage known as **Ohm's Law** in **Electrical Engineering**, to the relationship between Power, Temperature, and Heat Flow known as **Fourier's Law** in **Thermodynamics**, to the relationship between Power, Pressure, and Fluid Flow known as **Poiseuille's Law** in **Fluid Dynamics**, to the relationship between Power, Force, and Velocity known as **Dashpot** in **Mechanical Engineering**, all have a relationship to the Djed, Ankh, and Waas symbols representing stability, life, and power, held by the Ancient African deity of Engineering named **PTAH**. Therefore, the symbols of the Djed, Ankh, and Waas can be utilized as a **MASTER KEY** by Science and Engineering teachers, instructors, and professors to students of Egyptian, Kemetic, or African-centered studies, as part of a Transformative learning and Constructivist pedagogy.

It is quite remarkable that **Ptah**, the Ancient African deity associated with Engineering and Technology, would be holding a set of symbols which have such a synchronous association to symbols and concepts across electrical, thermal, fluid dynamic, and mechanical engineering and technology disciplines. All of these engineering disciplines come together to build an aircraft: **Electronics** gives you **Avionics** for Navigation and Control of the Aircraft, **Thermodynamics** gives you the science for the **Gas Turbine Engines** for **Thrust** of the Aircraft, **Fluid Dynamics** gives you the science of **Aerodynamic Lift** of the aircraft, and **Mechanics** gives you the science to **design**, **build**, **assemble**, **maintain**, and **repair** the aircraft. Moreover, the movement from the electron (electricity), to heat flow (thermodynamics), to element flow (hydraulics), to the movement of matter (mechanics) is consistent with the modern scientific description of the way sound is generated up from the movement of electrons to the movement of matter, and is also consistent with the cosmology found in the Memphite Theology of Ptah rising from the primordial abyss (electrical and thermal) and then imagining with his heart (hydraulic) and speaking with his tongue (mechanical) to produce sound. Indeed, the Djed, Ankh, and Waas are The Master Key mathematic symbols to African Science, Technology, and Engineering. A master key is One Key which can open up several different locks, and a master symbol is One symbol which can be used to apply to several different concepts. This is that key which will open the domain of multiple disciplines; for the key does not open the door, you do.

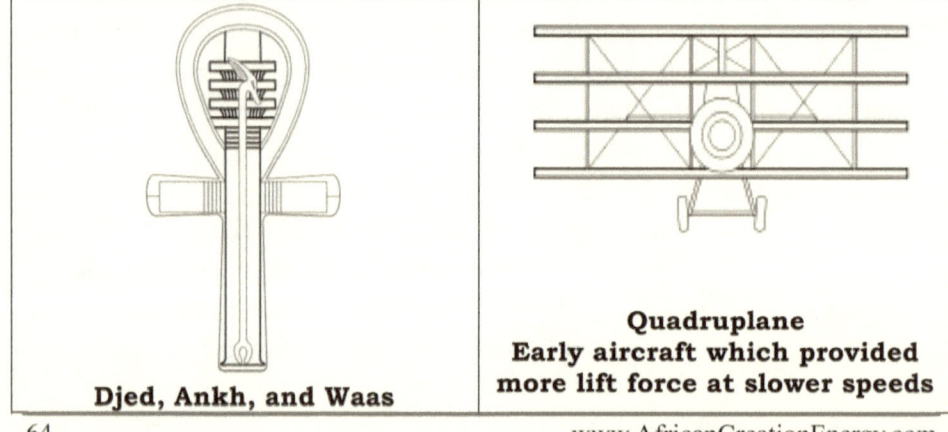

Djed, Ankh, and Waas

Quadruplane
Early aircraft which provided more lift force at slower speeds

Interdisciplinary Engineering Analogies

MASTER KEY		ELECTRICAL		THERMAL		HYDRAULIC		MECHANICAL
WAAS (Power)	**POWER (P)**	$P = I \times V$ $P = R \times I^2$ $P = V^2 \div R$	**POWER (P)**	$P = q \times T$ $P = R \times q^2$ $P = T^2 \div R$	**POWER (P)**	$P = G \times p$ $P = R \times G^2$ $P = p^2 \div R$	**POWER (P)**	$P = v \times F$ $P = R \times v^2$ $P = F^2 \div R$
ANKH (Life)	**CURRENT (I)**	$I = V \div R$ $I = P \div V$ $I = \sqrt{(P \div R)}$	**HEAT FLOW (q)**	$q = T \div R$ $q = P \div T$ $q = \sqrt{(P \div R)}$	**FLUID FLOW (G)**	$G = p \div R$ $G = P \div p$ $G = \sqrt{(P \div R)}$	**VELOCITY (v)**	$v = F \div R$ $v = P \div F$ $v = \sqrt{(P \div R)}$
DJED (Stability)	**VOLTAGE (V)**	$V = I \times R$ $V = P \div I$ $V = \sqrt{(P \times R)}$	**TEMPERATURE (T)**	$T = q \times R$ $T = P \div q$ $T = \sqrt{(P \times R)}$	**PRESSURE (p)**	$p = G \times R$ $p = P \div G$ $p = \sqrt{(P \times R)}$	**FORCE (F)**	$F = v \times R$ $F = P \div v$ $F = \sqrt{(P \times R)}$
PTAH	**RESISTANCE (R)**	$R = P \div I^2$ $R = V^2 \div P$ $R = V \div I$	**THERMAL RESISTANCE (R)**	$R = P \div q^2$ $R = T^2 \div P$ $R = T \div q$	**FLOW RESTRICTION (R)**	$R = P \div G^2$ $R = p^2 \div P$ $R = p \div G$	**FRICTION (R)**	$R = P \div v^2$ $R = F^2 \div P$ $R = F \div v$

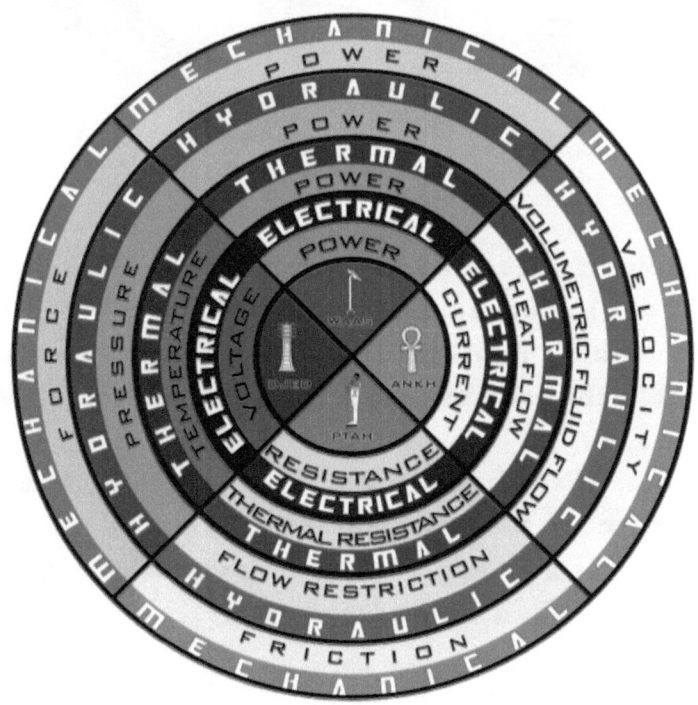

PTAH'S LAW OF ELECTRONICS

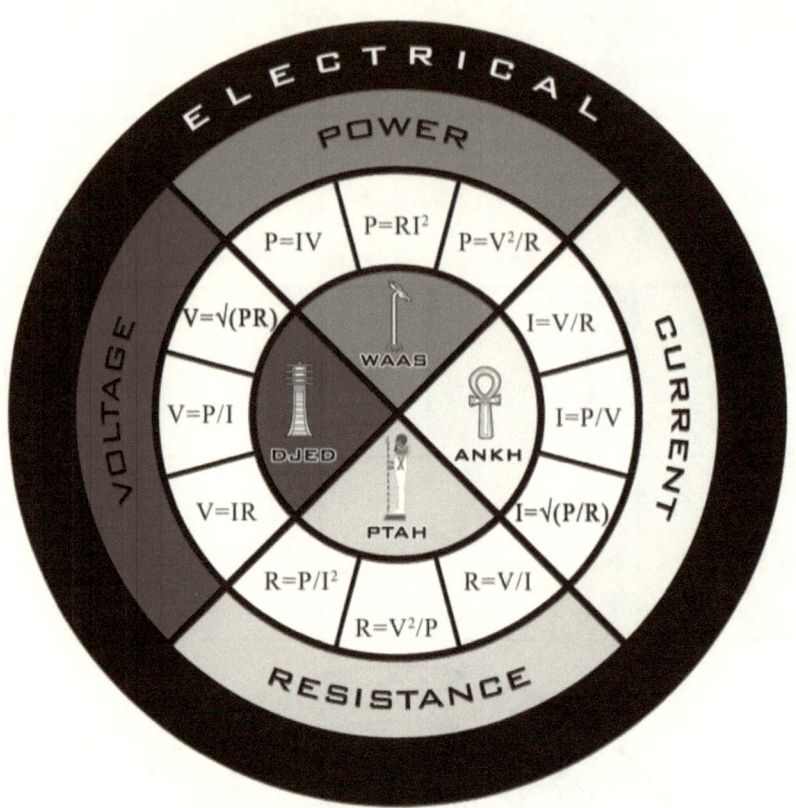

"Ptah's Law of Electronics" (known in modern science as "*Ohm's Law*") showing modern mathematic equations for the Electronic relationships and symbolism of the components that combine to form "staff of Ptah" circuit.

Djed, Ankh, Waas, and Ptah are represented with the modern mathematical symbols V, I, P, and R representing Voltage, Current, Power, and Resistance measured in Volts, Amperes, Watts, and Ohms respectively.

PTAH'S LAW OF THERMODYNAMICS

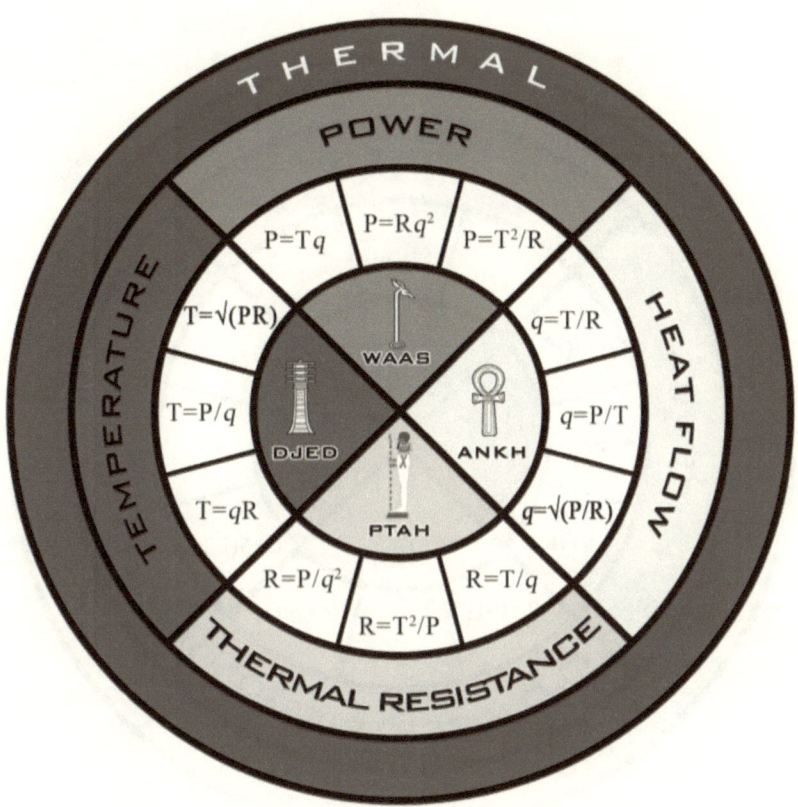

"Ptah's Law of Thermodynamics" (known in modern science as "*Fourier's Law*") showing modern mathematic equations for Thermodynamics relationships and symbolism of the components that combine to form "staff of Ptah".

Djed, Ankh, Waas, and Ptah are represented with the modern mathematical symbols T, *q*, P, and R representing Temperature, Heat Flow, Power, and Thermal Resistance respectively.

PTAH'S LAW OF FLUID MECHANICS
(Aerodynamics & Hydrodynamics)

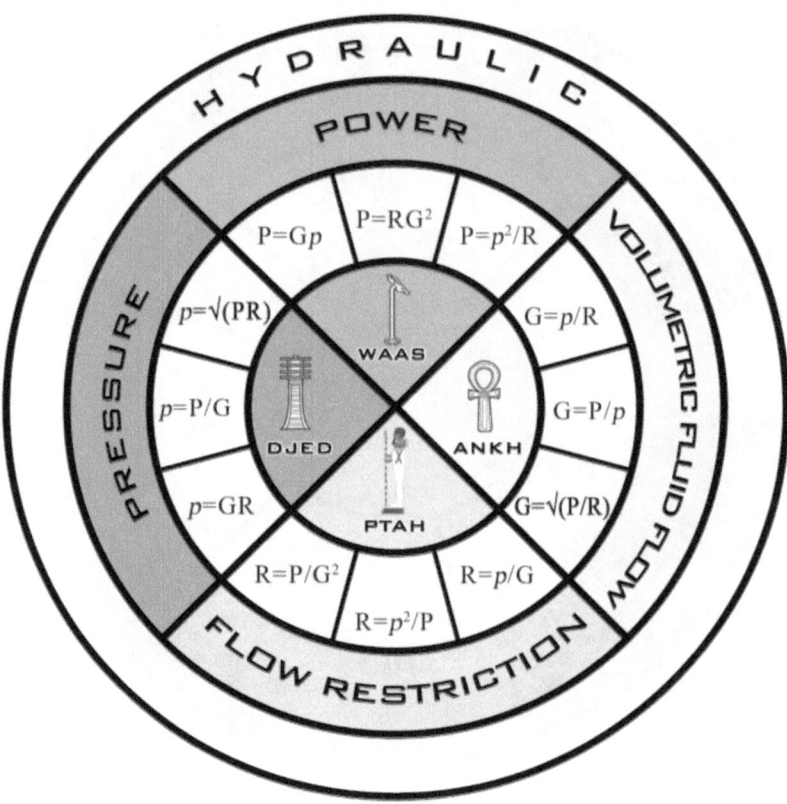

"Ptah's Law of Fluid Dynamics" (known in modern science as *"Poiseuille's Law"*) showing modern mathematic equations for Fluid Dynamic relationships and symbolism of the components that combine to form "staff of Ptah".

Djed, Ankh, Waas, and Ptah are represented with the modern mathematical symbols p, G, P, and R representing Pressure, Fluid Flow, Power, and Flow Restriction respectively.

PTAH'S LAW OF MECHANICS

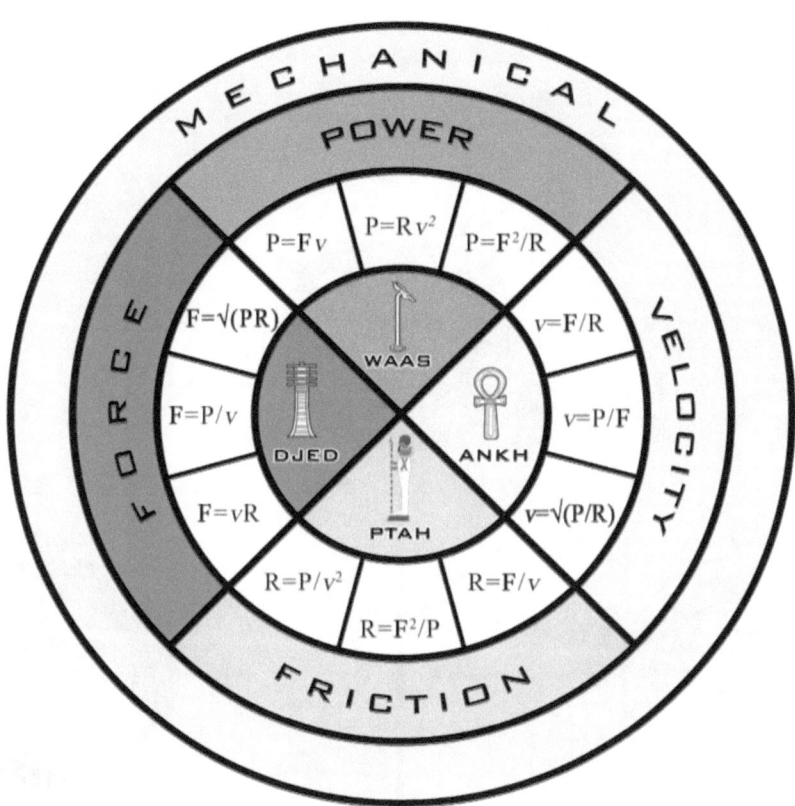

"Ptah's Law of Mechanics" (known in modern science as "*Dashpot*") showing modern mathematic equations for Fluid Dynamic relationships and symbolism of the components that combine to form "staff of Ptah".

Djed, Ankh, Waas, and Ptah are represented with the modern mathematical symbols F, *v*, P, and R representing Force, velocity, Power, and Friction respectively.

4 Engineering Sciences needed for Aeronautical Ascension:

- Electrical Engineering
- Thermodynamics
- Aerodynamics
- Mechanical Engineering

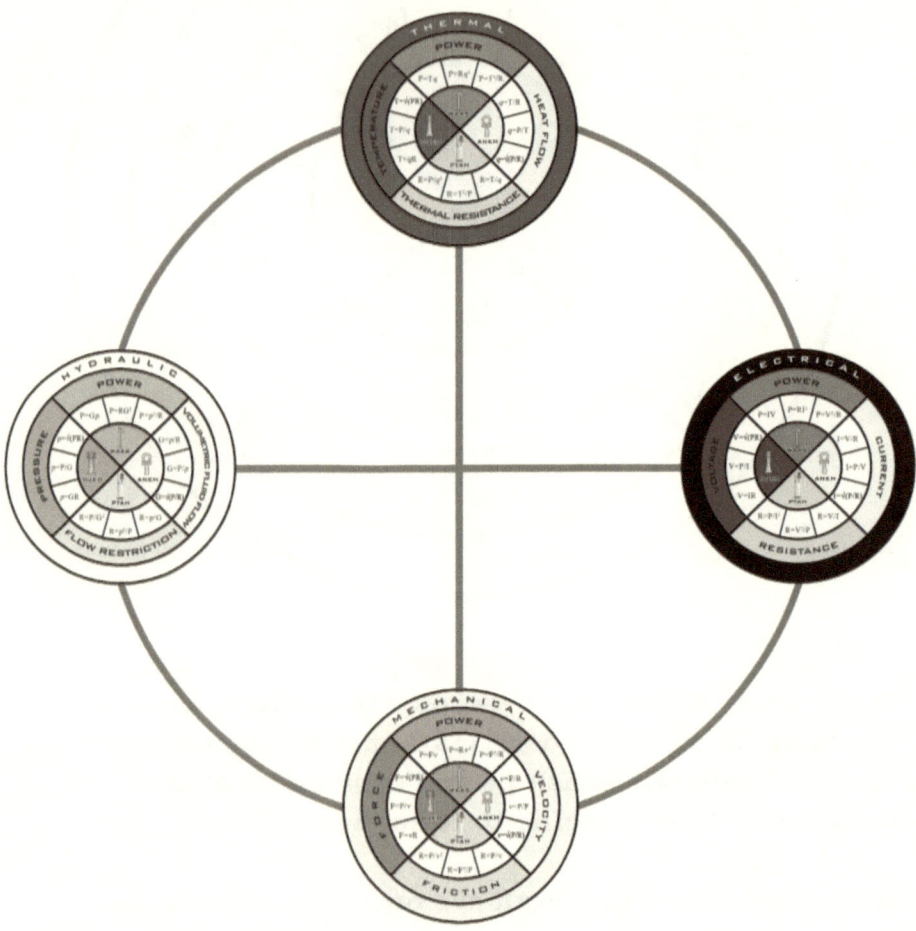

4. Aerodyne Aircraft

In Aerodynamics, Aeronautics, and Aviation, the term **"Aerodyne"** is used to refer to any aircraft which utilizes **Aerodynamic Lift** for flight as opposed to **buoyancy**. Aircraft which use the principle of buoyancy to achieve flight are called Aerostats and are considered "lighter-than-air" aircraft; these include aircrafts like hot-air balloons, blimps, airships, and zeppelins. Aerodyne Aircraft are considered "heavier-than-air" aircraft. Since Aerodyne Aircraft use Aerodynamic Lift to achieve flight, which is in turn determined by the wing or airfoil of the aircraft, then it makes sense that Aerodyne Aircraft can be generally classified by their wings being either 1) movable wing aircraft or 2) fixed wing aircraft.

Under the movable wing category of aerodyne aircraft, there are two sub-groups: 1) Ornithopter Aircraft and 2) Rotary Wing Aircraft. Ornithopter aircraft is the aircraft design most favored by Nature. The word "Ornithopter" comes from the Greek words "Ornitho-" meaning "bird" and "-pteron" meaning wing, and refers to aircraft which fly using wings which have a flapping motion like birds and insects. Most human created aircraft do not follow the ornithopter design, and the human created aircraft which do follow the ornithopter design have been either aircraft designed for a single person (like a hang glider), or un-manned remotely controlled or autonomous aircraft created by hobbyists.

Rotary wing aircraft include aircraft which utilize propellers primarily for lift like helicopters. Technically, Turbofans, Turboprops, Turbojets, and Jet Engines on Fixed-wing aircraft all have propellers, rotors, and turbines. However, what distinguishes Rotary wing craft like helicopters from fixed-wing aircraft which also have propellers is that the propellers on rotary wing aircraft are oriented to give the aircraft lift, whereas the propellers on fixed-wing aircraft are oriented to give the aircraft thrust. The etymology of the word "**Helicopter**" means "**Rotating Wing**" (*Revolutionary Uplift*) coming from the Greek word "Heli" meaning "helix, spiral, turning, twisting, revolving" and the Greek word "-pteron" meaning "wing". It is the fact that all Aerodyne aircraft utilize a propeller or a form of "Rotating Wing" (Revolutionary Uplift) which is the etymological meaning of the word "Helicopter" and is why this book has been entitled "*H.E.R.U.-copters*"; a play on the word "Helicopter". The "H.E.R.U." acronym in the word "*H.E.R.U.-copters*" stands for "Helicopter Enabling Rise and Uplift". Rotary wing aircraft, or "Helicopters", are classified by the number of propellers the aircraft has. A Helicopter with 1 propeller can be considered a **Mono-copter**, and would be similar to a **U.F.O. Flying disc** design. A Helicopter with 2 propellers can be considered a Bi-copter, and would include most of the Helicopter designs people are familiar with today which have 2 propellers – a main propeller and a tail propeller. Additionally, "personal helicopters" in the form of light-weight one-passenger aircraft, or "helicopter backpacks" have been built based on the Mono-copter and Bi-copter design.

Bi-copter aircraft called **Auto-gyros** have 2 propellers, 1 powered propeller in the front to provide thrust, and 1 un-powered propeller on the top which automatically turns due to air moving up through the rotor to provide lift. Helicopters with 3 propellers (Tri-copters), 4 propellers (Quadcopters or Quadrotors), and Multiple Propellers (Multi-copters) have become increasingly popular designs for "drones" or autonomous, semi-autonomous, and remote-controlled aircraft due to the advances in microchip technology which provides a platform, method, and means to control and quickly balance all of the propellers of Multi-propeller aircraft. Since we live in a 3-Dimentional world, aircraft have 3 possible **Degrees of Freedom of Movement** which they could potentially maneuver in: 1) up-down, 2) left-right, and 3) backwards-forwards. Thus, at the very minimum, Multi-propeller Helicopters need 3 propellers, or a Tri-copter configuration, to maximize their utilization of potential Degrees of Freedom of Movement. Quad-copters or Quad-rotor configurations with 4 propellers, and Multi-copters with more than 4 propellers, provide even more control over the aircraft's degrees of freedom of movement. A multi-copter's ability to make use of all of the available degrees of freedom of movement makes it a more aerobatic aircraft capable of performing more complicated maneuvers than traditional 2-rotor Helicopters. Also, every propeller over 3 that a Multi-rotor Helicopter has provides the aircraft with **Redundancy** so that in the event one of the propellers stops working, there are still at least 3 propellers available for the Multi-copter Helicopter to fly.

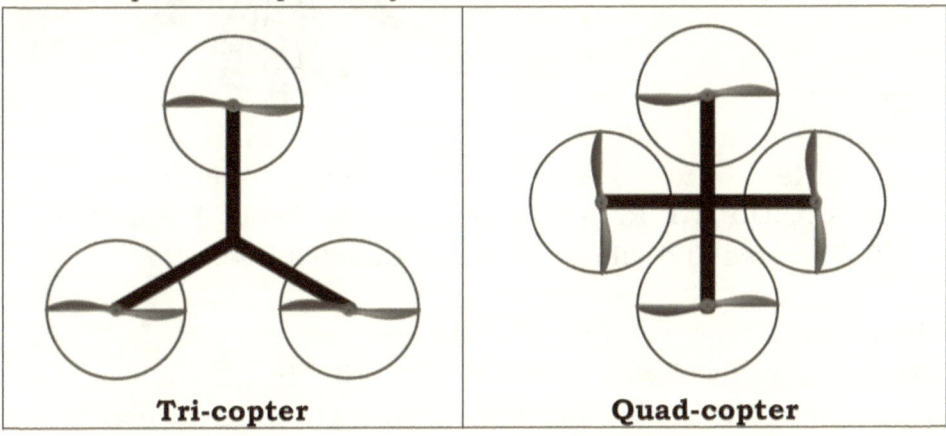

Tri-copter   **Quad-copter**

The use of Multi-copters for manned flights is still relatively new. The world's first manned flight with an electric Multi-copter occurred on October 21, 2011 in an aircraft called the **Volocopter VC1**. The Tri-copter configuration may also be implemented in the long rumored **"Black triangle U.F.O."** called the **Astra Aurora TR-3B** which is said by

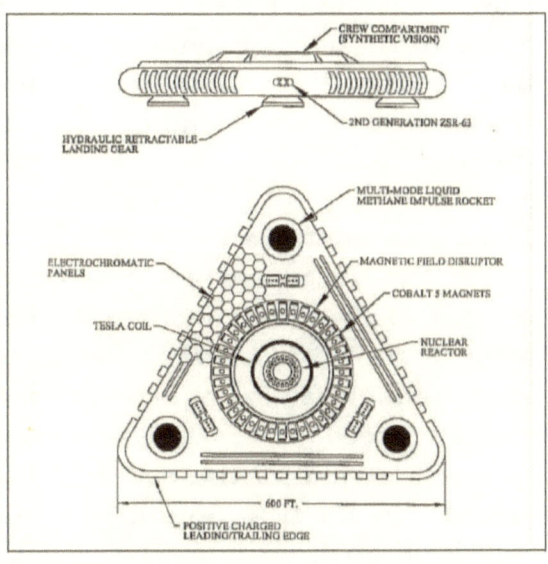

Ufologists to be a large triangular-shaped flying machine capable of vertical lift and silently hovering in the sky at low altitudes.

The **Springtail EFV-4** is a Bi-rotor personal helicopter backpack capable of carrying a single person 1900 meters in the air at a top speed of 100 km/h. A 3 propeller Tri-rotor design can also be used based on the 3 spheres found in some depictions of the Ancient African Aviation

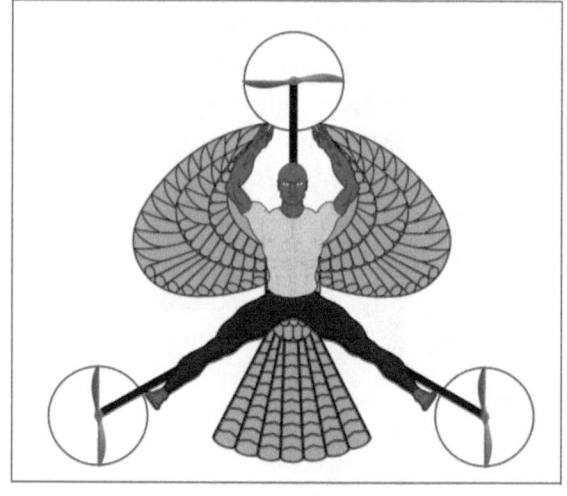

deity Heru to create a personal helicopter or **H.E.R.U.C.O.P.T.E.R.** (Helicopter Exoskeleton Rotor Uplift Craft Operated and Piloted by Three Enabled Rotors).

While Ornithopters and Helicopters are not as fast and aerobatic as fixed-wing aircraft, the primary advantages that Helicopters have over Fixed-wing aircraft are the ability to Hover and the ability to Vertically Take-off and Land (VTOL). Some of the most popular Helicopters are listed on the next page. The U.S. Military Helicopters are commonly given "nicknames" corresponding to **Native American Indian** themes.

Fixed wing aircraft are exactly what the name suggests, aircraft where the wings are not required to move in order for the aircraft to fly. Most modern commercial airplanes are fixed-wing aircraft. Fixed-wing Jet aircraft are able to travel at speeds much faster than Helicopter and rotary wing aircraft. Aircraft speed is generally characterized by **Mach** number, which is a ratio of the speed of the aircraft to the **speed of sound**. In Physics, **Sound** is a vibration that propagates or travels through mediums like solids, liquids, and air. Sound can be audible or inaudible.

Helicopter	Capability
UH-60 Black Hawk Manufacturer: Sikorsky	utility helicopter used for attack, cargo, and evacuation
AH-64 Apache Long bow Manufacturer: Boeing	Gunship / attack helicopter equipped with cannons, machine guns, rockets, hellfire missiles, air-to-ground missiles, and air-to-air missiles
RAH-66 Comanche Manufacturer: Boeing-Sikorsky	Stealth Helicopter
X^3 Manufacturer: Eurocopter	World's fastest Helicopter as of June 2013 capable of 300 mph

The **speed of sound** traveling through air is **323 meters per second** or **767 miles per hour**. When an aircraft accelerates from below the speed of sound to above the speed of sound, a shockwave which sounds like an explosion is created called a **Sonic Boom**. The Mach number that an aircraft is capable of traveling at is determined by the type of propulsion system that it uses for thrust. The various Mach numbers are categorized in the table below:

Mach Number	Description	Speed
Mach <0.8	Sub-sonic	<610 mph
Mach 0.8 – 1.2	Transonic	610 – 915 mph
Mach 1	Sonic	767 mph
Mach 1.2 – 5.0	Supersonic	915 – 3,840 mph
Mach 5.0 – 10.0	Hypersonic	3,840 – 7,680 mph
Mach 10.0 – 25.0	High Hypersonic	7,680 – 16,250 mph
Mach 25.0 – 881,000	Space Travel	>16,250 mph
Mach 881,000	Speed of Light	671,000,000 mph

Aircraft which get their thrust from propellers are generally slower than aircraft which get their thrust from **Jet Engines** and **Jet propulsion.**

Gas **Turbine Engines** are the most common type of propulsion system used on aircraft. Gas Turbine Engines come in 4 forms: 1) **Turboshaft** – a turbine engine

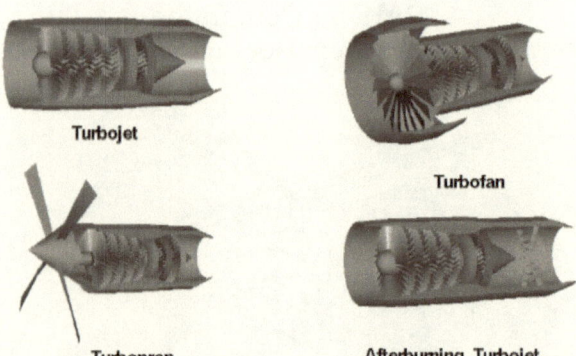

Turbojet

Turbofan

Turboprop

Afterburning Turbojet

designed to provide more power than thrust and used for slow subsonic flying aircraft like Helicopters; 2) **Turboprop** – a propeller driven by a turbine engine and used for subsonic flight speeds; 3) Turbofan – a shrouded or ducted propeller (which reduces thrust loss) driven by a turbine jet engine and used for Transonic flight speeds; 4) **Turbojet** – a series of ducted propellers driven by a turbine engine which take in air, compress the air, then burn and combust the air to produce thrust to travel at supersonic flight speeds. The Turbojet engine can be modified with **Afterburner** to travel from Supersonic to Hypersonic flight speeds.

One of the most notable **Supersonic** aircraft is the **SR-71 Blackbird Habu** (manufactured by Lockheed) which currently holds the

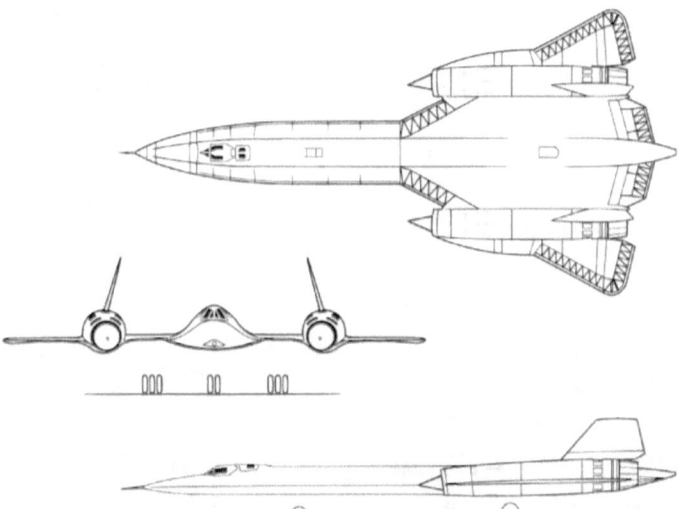

record for the world's fastest air-breathing manned aircraft. The top speed of the SR-71 Blackbird Habu is top secret, but it can reportedly travel at speeds of Mach 3 and greater. The SR-71 Blackbird Habu "*raced the Sun and won*" which means, since the rising and setting of the Sun as seen from the planet Earth is determined by the rotational speed of the Planet Earth, and the circumference of the Planet Earth is 25,000 miles at the Earth's equator, and the Earth makes 1 revolution every 24 hours, then in order for you to race the Sun across the sky and win the race, you would have to travel at a speed greater than 25,000 miles÷24 hours = 1,042 miles per hour (≈ Mach 1.5). An **Unmanned Aerial Vehicle (UAV)** Hypersonic version of the SR-71 Blackbird Habu named the **SR-72** capable of Mach 6 and greater was announced to be in development in November 2013.

Anecdotally it is said that **Clarence Kelly Johnson**, team leader of the Lockheed Skunk Works and designer of the SR-71 Blackbird Habu, reported seeing a "***crescent shaped***" U.F.O. 13 years prior to designing the SR-71 Blackbird Habu. Lockheed, the same company which manufactured the SR-71 Blackbird Habu and the SR-72, also manufactured the first aircraft designed with **stealth technology** named the **F-117 Nighthawk**. The use of stealth technology on aircraft was improved upon when Northrop

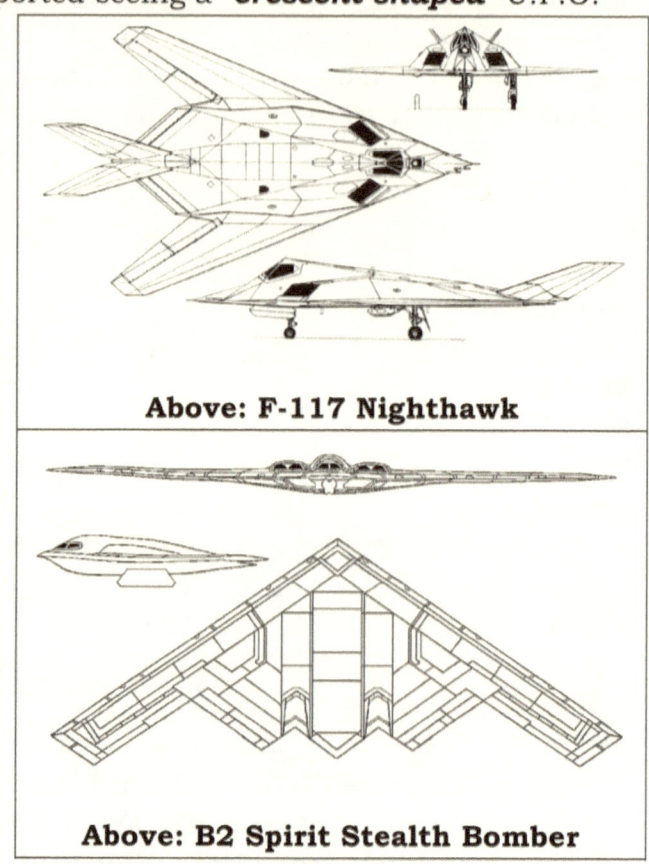

Above: F-117 Nighthawk

Above: B2 Spirit Stealth Bomber

Grumman developed the **B2 Spirit Stealth Bomber** which uses the "**flying wing**" aircraft design. A "Flying Wing" aircraft design is a tailless aircraft where the body and fuselage are housed within the wing of the aircraft. Flying wing aircraft tend to be "***crescent shaped***". One of the earliest flying wing aircraft was the **Horten Ho 229** designed by the German Horten brothers in 1944. Later, the first time the term "**Flying Saucer**" was used by the media was to describe Unidentified Flying Objects (U.F.O.s) seen by Kenneth Arnold near Mount Rainier, Washington on June 24, 1947 which he actually described as **9 Flying Crescent-Shaped aircraft**.

Aircraft can also be designed with **Variable Geometry** to optimize performance in certain circumstances. For fixed-wing aircraft, the "**Swing Wing**" variable geometry configuration allows the aircraft to have its wings swept back, like flying-wing aircraft, during high-speed 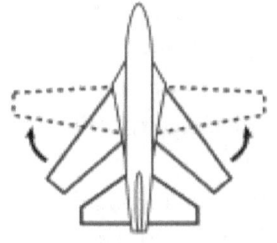 flight, and the wings can be brought back to the original position during lower-speed flights. The variable sweep wing gives the pilot the ability to choose the appropriate wing configuration given the aircraft's speed much like a "stick-shift" manual transmission car allows the driver to select the appropriate gear for the car's speed.

The **Tiltrotor** aircraft is an example of a Helicopter or Rotor aircraft which utilizes a variable geometry configuration. The propellers of the Tiltrotor aircraft are parallel with the ground during Vertical Take-Off and Landing (VTOL) and as the aircraft gains speed, the propellers are tilted forward to give the aircraft forward thrust. There are advantages and disadvantages to the fixed wing and rotary 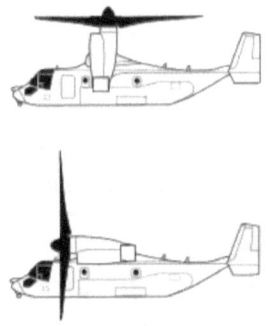 wing aircraft design. Fixed wing aircraft are capable of aerobatic maneuvers and are able to travel at faster speeds; some fixed-wing aircraft are capable of flying faster than the speed of sound. Rotary wing aircraft, which include "flying saucers", are capable of hovering and vertical take-off and landing (VTOL). **Tiltrotor** and **Tiltwing** aircraft are attempts to blend the Vertical Take-Off and landing and hovering characteristics of rotary wing aircraft with the speed of fixed-wing aircraft. Recall that "flying saucers" and aircraft capable of Vertical Take-Off and landing and hovering tend to have "Circular" (O) geometries, and aircraft capable of high-speeds and aerobatic maneuvers have "Cross-shaped" (+) geometries, then in aircraft, the unification of the principles of Vertical Take-Off and landing and hovering (**O**) and high-speed and aerobatic maneuvers would give the **Ankh** (♀); a symbol of the unification of dual principles. The blending of Vertical Take-Off and landing, hovering, supersonic speed, and aerobatic maneuverability characteristics into a single aircraft has yielded what is called **Air Superiority** and **Air Supremacy Vehicles**. The first Supersonic Jet capable of Vertical Take-Off and landing and hovering was the **Harrier Jump Jet**. Vertical Take-Off and landing capabilities have also been implemented in the Lockheed Martin **F-22 Raptor** and **F-35 Lightning** supersonic Air Superiority vehicles.

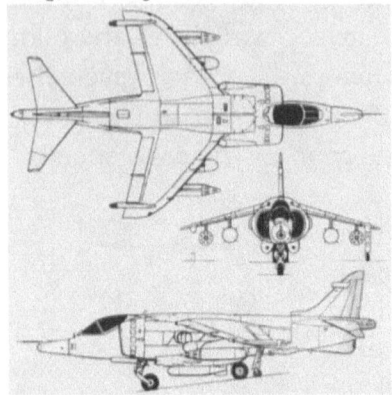

Harrier Jump Jet

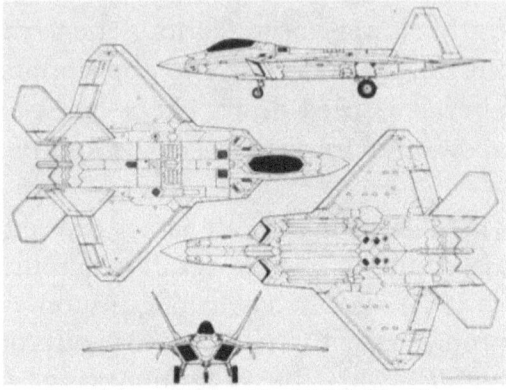

Lockheed Martin F-22 Raptor

Air Superiority Supersonic Jets capable of Vertical Take-Off and landing utilize a **Liftfan** or **Liftjet**, which is a jet which provides a downward thrust to lift the aircraft upward, along with **Thrust Vectoring**, which is the ability to control the

Lockheed Martin F-35 Lightning

direction of thrust, to enable the aircraft to take-off vertically, hover, and perform a variety of aerobatic maneuvers.

As flight speeds increase, and the aerobatic maneuvers of aircraft become more complex, the **G-Force** that human pilots experience while inside of the aircraft can become more than a Human-being can endure. However, **Unmanned Aerial Vehicles** (**UAV**), colloquially known as "***Drones***", are capable of faster aerial speeds and more complex aerobatic maneuvers since a human pilot is not present in the aircraft. The "***Rise of the Drones***" is expected to be the next **Air Superiority** and **Air Supremacy** Vehicle. Currently, and in the near future, ordinary people will be able to DIY (do-it-yourself) build, own, and operate drone aircraft and spacecraft, which has prompted the **FAA** (**Federal Aviation Administration**) to closely monitor and regulate the use of Drone

"**Guizhou Soar Dragon**" **Chinese military UAV**

technology. Drone aircraft have countless uses and application outside of military including reconnaissance, information gathering, package delivery, aerial video and photography, weather and traffic monitoring and reporting, disaster relief search and rescue and aid delivery, and agriculture. Humanity has gone from the Biblical, Religious, and Theological "**War In Heaven**" (**Battle in the Sky**) between the "Angel Michael and the Fallen Angels," to a literal "**War for Heaven**" using science, technology, and drone aircraft. The 7 names used to refer to Unmanned Aerial Vehicles or "*Drones*" are:

1. UAV - Unmanned Aerial Vehicle
2. UCAV - Unmanned Combat Aerial Vehicle
3. RPV - Remotely Piloted Vehicle
4. RPA - Remotely Piloted Aircraft
5. ROA - Remotely Operated Aircraft
6. UAS - Unmanned Aerial System
7. Drone

The **"Defense Advanced Research Project Agency"** (**DARPA**) has developed a Hypersonic UAV Drone aircraft capable of **Mach 20** speeds named the **HTV-2 Falcon**. As a comparison, **space shuttles** are capable of **Mach 22** speeds.

DARPA HTV-2 Falcon

BAE Systems has developed a **Stealth Supersonic drone UAV** aircraft called **Taranis**. In Celtic mythology, Taranis was a **sky deity** and **deity of thunder** who was depicted with a thunderbolt in one hand and a "wheel" in the other hand. The Celtic deity Taranis is connected with the Norse deity **Thor** and the Roman deity **Jupiter**. The book entitled **"P.T.A.H. Technology: Engineering Applications of African Sciences"** discusses how the Norse deity Thor, the Roman deity Jupiter, and the Celtic deity Taranis are derivatives of the African Creation deity named **PTAH**.

BAE Systems "Taranis" UAV Stealth Drone Aircraft

Celtic deity Taranis

Wheel of Taranis

Aircraft are tools of technology, and like any tool or technology, it can be used for constructive or destructive purposes depending on the intention of the person wielding the tool.

5. Aviation Dynamics:

Whereas the topic of "**Ascension**" can be thought of as the spiritual, metaphysical, and subjective "*wing*" of the discussion of flight, and the topic of "**Aerodynamics**" can be thought of as the scientific, theoretical, and objective "*wing*" of the discussion of flight, **Aviation** is the operative art of aeronautic flight; i.e. the actual flapping of the "*wings*". **Aviation Dynamics**, or **flight dynamics**, is the science of orienting, stabilizing, and controlling the position of an aircraft in the air.

While flying in the air, aircraft move in 3 Dimensions: 1) Forwards and Backwards, 2) Left and Right, and 3) Up and Down. On a mathematical 3-Dimensional coordinate system, these 3 dimensions can be represented by the x-axis, y-axis, and z-axis respectively. The 3 terms used in the field of Aviation Dynamics to discuss an aircraft's movement in these 3 dimensions are:
1. **Pitch** – a movement of the aircraft up or down. Pitch is a movement by an aircraft which changes the vertical direction the aircraft's nose is pointing.
2. **Yaw** – a movement left or right along. Yaw is a movement by an aircraft which changes the horizontal direction the aircraft's nose is pointing.
3. **Roll** – a diagonal movement. Roll, also called **Banking**, is a movement by an aircraft which changes the orientation of the aircraft's wings relative to the ground.

On **Fixed wing aircraft**, Pitch, Yaw, and Row Aviation Dynamics are controlled by the following mechanical systems:
1. **Elevators** – flaps usually located near the rear of the aircraft with control **Pitch**
2. **Rudder** –usually located on the tail of the aircraft which controls **Yaw**.
3. **Ailerons** – flaps usually located on the wing of the aircraft which control **Roll**.

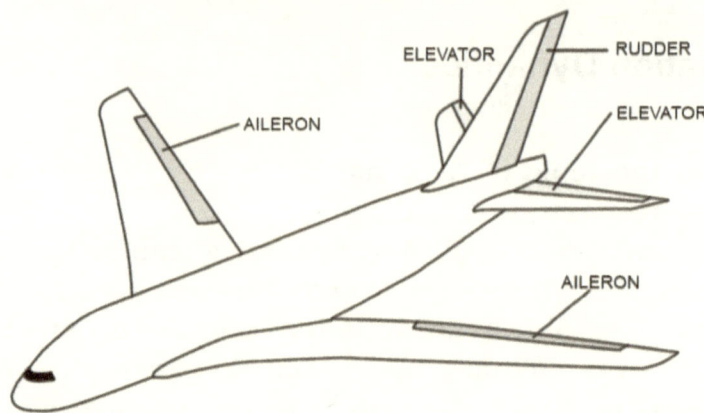

The airplanes Pitch and Roll, which are manipulated by the Elevators and ailerons, are controlled by the pilot's Yoke, or steering control column; the Yaw and Rudder is controlled by pedals. Airplanes can fly forward, pitch upward to climb, pitch downward to dive, and can turn by either yawing or rolling left or right. If an airplane wanted to go back the way it came, it could not fly backwards, but rather it would have to cover a large amount of airspace to turn to point its nose in the opposite direction. Conversely, the Aviation Dynamics of Helicopters and rotor craft are much more maneuverable.

Helicopters are capable of all the Aviation Dynamic maneuvers of an

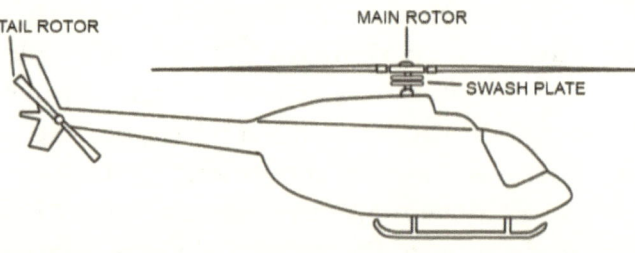

Airplane, but also Helicopters can fly backwards, side-to-side, rotate in one spot, and hover. On **Helicopters**, Pitch, Yaw, and Row Aviation Dynamics are determined by the orientation of the helicopter's main rotor and are controlled by the mechanical system called the **Swash Plate** which is located underneath the helicopter's main rotor. Basically, the direction in which the main rotor is tilted will determine the direction the helicopter will fly, and the swash plate is what tilts a helicopter's main rotor. The swash plate of a helicopter is controlled by the helicopter pilot using the **Cyclic Lever** and the **Collective Lever**. The Collective Lever maneuvers the helicopter up-and-down, while the Cyclic Lever maneuvers the helicopter forward, backward, and side-to-side. The tail rotor is controlled by the helicopter pilot's pedals, and the tail rotor controls the Yaw, or left and right, flight dynamics of the helicopter.

The Pitch, Yaw, and Roll Aviation Dynamics for Multi-rotor aircraft are controlled by balancing the power given to 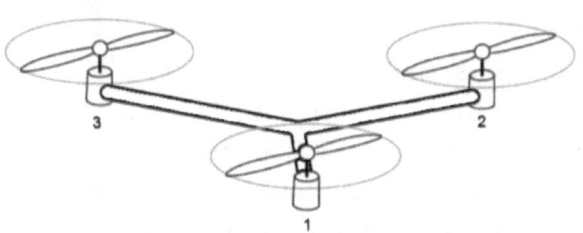 each of the aircraft's rotors. On **Tri-copters**, aviation dynamics are controlled by the mechanical systems which make up the 3 rotors of the Tri-copter. In the image of the Tri-copter above, imagine that rotor 1 and rotor 2 are the two front rotors of the Tri-Copter, and rotor 3 is the rear rotor of the tri-copter. If rotor 1 is designed to spin clockwise, then rotor 2 must be designed to spin counter-clockwise to offset the torque. Rotor 3 will then have to be designed to spin either clockwise or counter-clockwise. Whichever direction is chosen for rotor 3 to spin, either clockwise or counter-clockwise, the Tri-rotor will have a tendency to Yaw in that direction, and thus the rear rotor of the Tri-copter, must be designed to be able to tilt left and right to offset the Yawing due to 2 of the 3 Tri-copter rotors spinning in the same direction. For Tri-copters, the Yaw and pitch flight dynamic is controlled by tilting a rotor (in our example rotor 3) left or right. With all 3 rotors pointing up, the tri-copter can vertically take-off and hover. Changing the power distribution to the rotors of the tri-copter will cause the aircraft to tilt and fly in that direction. The aviation dynamics for Tri-rotors is unique amongst multi-rotor aircraft because tri-rotors require at least one of the rotors to be able to tilt, whereas multi-rotor aircraft with 4 or more rotors are able to control Pitch, Yaw, and Row Aviation dynamics by only controlling the amount of thrust provided by each of the rotors.

Imagine that in the image of a **Quad-copter** to the right, rotor 1 is the front rotor and rotor 3 is the 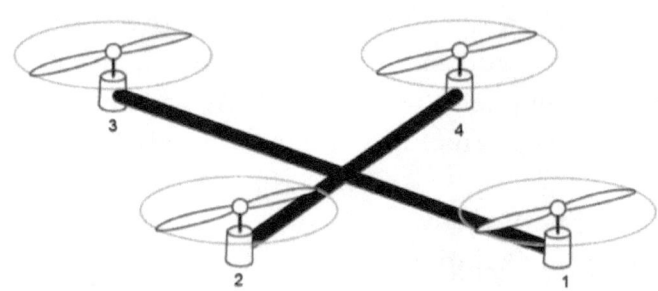 back rotor. To offset the torque forces, if rotors 1 and 3 are designed to spin clockwise, then rotors 2 and 4 should be designed to spin counter-clockwise. When equal thrust is applied to all four rotors, then the Quad-copter aircraft is capable of ascending, adjusting in altitude, and/or hovering. If more thrust is given to rotors 1 and 3 which are spinning clockwise, then the Quad-copter would Yaw in the clockwise direction. If more thrust is given to rotors 2 and 4 which are spinning counter-clockwise, then the Quad-copter would Yaw in the counter-clockwise direction. If more thrust is given to rotor 1 than rotor 3, then the Quad-copter would pitch upward. If more thrust is given to rotor 3 than rotor 1, then the quad-copter would pitch downward. If more thrust is given to rotor 2 than rotor 4, then the quad-copter would roll in the direction of rotor 2. If more thrust is given to rotor 4 than rotor 2, then the quad-copter would roll in the direction of rotor 4.

Aviation Dynamics are controlled by a combination of mechanical and electrical systems. The ailerons, elevators, rudders, swash plates, and rotors discussed here are all the mechanical systems which go into Aviation Dynamics. The next section on Avionics discusses the electrical systems which enable an aircraft to fly. "Aviation Dynamics" is simply the direction which a flying object traverses in the air. With Heru as our African deity of Aviation, we can think of **"Aviation Dynamics"** as **"Heru's Journey"**. In literature, the **"Hero's Journey"** is a common theme found in various stories about Heroes. The "Hero's Journey is usually divided into 4 main parts: 1) The Call, 2) Descent, 3) Transformation, and 4) The Return. Since the word "Hero" is phonetically and etymologically similar to the name Heru, then we will redefine the **"Hero's Journey"** of literature to reflect **"Heru's Journey"** as an **Aviator** using **Flight Dynamics** to consist of the following 4 main parts: 1) Call to Fly, 2) Ascension to Heaven, 3) Transcendence, and 4) Return.

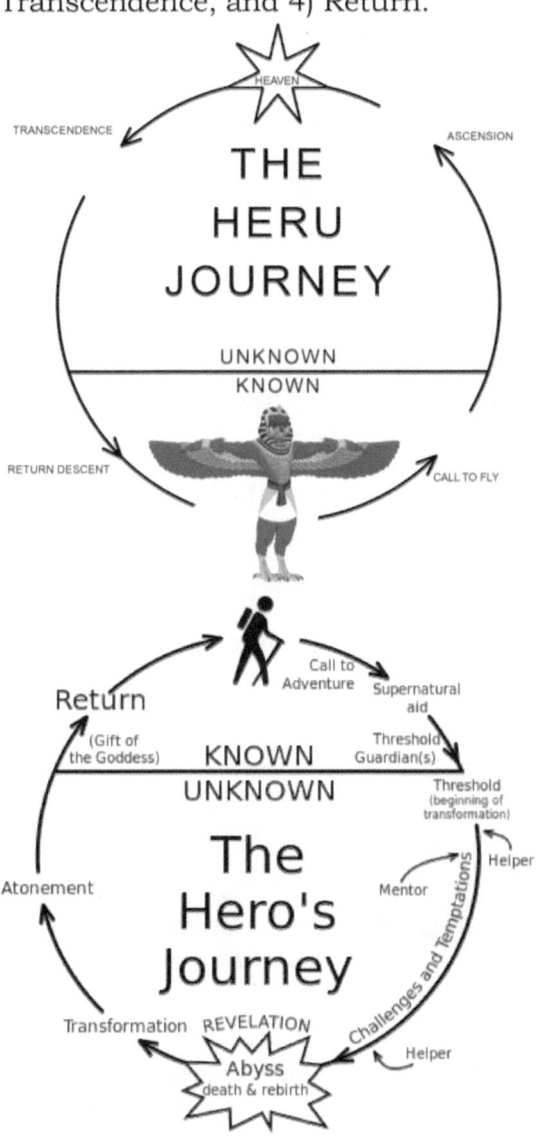

6. Avionics

The word "**Avionics**" refers to the Aviation Electronics used on board an aircraft. Avionics includes all the electrical instruments, equipment, and computer software used by the pilot to control and monitor flight control and stability, navigation, and communication. Examples of Avionic equipment include autopilot systems, anti-collision systems, and radar. While the sensors and wiring for Avionic equipment is located throughout the aircraft, the Avionic user interfaces are located in the flight deck of the aircraft to be monitored by the aircraft's pilot. Some of the most critical Avionic flight instruments include:

ASI   **Airspeed Indicator** – indicates the speed the aircraft is travelling

AI   **Attitude Indicator – (Gyro Horizon Indicator)** - indicates the orientation (pitch and roll) of the aircraft relative to the Earth's Horizon. If the aircraft is not level relative to the Horizon, it is considered a "*Bad Attitude*".

ALT   **Altimeter (Altitude Meter)** - indicates the aircraft's altitude above sea-level

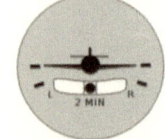

TC   **Turn Coordinator (Inclinometer)** - indicates the rate of roll, or turn, for the aircraft

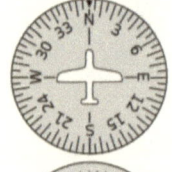

DG   **Directional Gyro (Heading Indicator)** - the aircraft's heading with respect to magnetic north

VSI   **Vertical Speed Indicator (Variometer)** – indicates the Pitch, rate of Ascent (climb) or rate of descent of the aircraft

Originally many flight instruments were mechanical and analog, but in modern times electrical and digital flight instruments have become more popular. The Attitude Indicator, Turn Coordinator, and Directional Gyro (Heading Indicator) flight instruments are all **Gyroscopic sensors**. A Gyroscope is an instrument that uses angular momentum to measure and maintain orientation, balance, and stability. Gyroscopic sensors are also used within cell phones to detect horizontal and vertical orientation.

Mechanical Gyroscopes are designed as a **wheel-within-a-wheel** that spins along an axis and can maintain balance while spinning; much like a **spinning-top**. However, there are no spinning top devices inside cell phones and airplanes which perpetually spin to measure and maintain orientation.

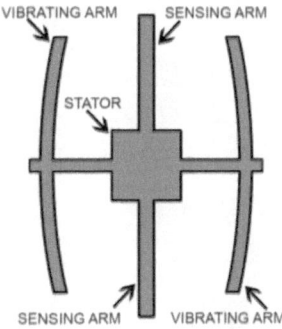

Gyroscopic Sensors work on the principle of **Piezoelectricity** in **Micro Electro-Mechanical Systems (MEMS)**. Inside of a Gyroscopic sensor IC (Integrated Circuit) there is an **"H"** shaped quartz crystal structure. When an electric current is applied to the H-shaped structure, the arms of the "H" begin to vibrate due to the piezoelectric effect. If the orientation of the "H" is changed, the arms of the H will move in a different direction, and this change can be detected by the piezoelectric signal it creates from the "H" structure to the IC. The word "Gyroscope" comes from the Greek word "gyro" meaning circle. The word "gyro" in Greek is also pronounced "*Yeer-ro*" or "*Heer-ro*" referring to "circular" sandwiches. So it is interesting to note the phonetic similarity between the African Aviation deity Heru and the Avionic Instrument Gyro (Heer-Ro) used to maintain a level Horizon. Also, the word "**Horizon**" is phonetically similar to the words **HERU RISING**. In order to Experience Evidence of the Avionic systems which go into the design of an aircraft, the next chapter on Aeronautical Engineering provides the reader with a project to build a UAV Drone aircraft.

7. Aeronautical Engineering Experiments

Aeronautical Engineering is the branch of engineering which deals with the research, design, development, construction, testing, science and technology of aircraft which operate in Earth's atmosphere. Aeronautical Engineering is combined with **Astro-nautical Engineering** (the engineering of spacecraft) to form the field of **Aerospace Engineering**. The following Aeronautical Engineering Experiment provides instructions on how to build a "**Heru-Copter Tri-Rotor Aircraft**".

Materials Needed:

Quantity	Description
2	Plywood (10cm wide x 15cm high x 0.3cm thick)
3	pinewood (1.3cm wide x 1.3cm high x 50cm long)
1	pinewood (1.3cm wide x 1.3cm high x 5cm long)
1	pinewood (1.3cm wide x 1.3cm high x 4cm long)
1	pinewood (1.3cm wide x 1.3cm high x 3.5cm long)
3	18 - 20 amp ESCs (Electronic Speed Controllers - Turnigy Plush)
3	Brushless Motors (DT750 750kV)
3	Propellers (GWS 11*4.7)
1	Battery (3s Turnigy 25-35C 2200mAh LiPo)
1	Servo (BMS-385DMAX Digital Servo - Metal Gear)
1	Bottle of Blue Locktite
1	Bottle of wood glue
11	#6 -32 x 1 inch Flat Head Phillips screws, nuts, and washers
1	Pack of 100 Zip Ties
12	12 pair (24 total) of Male-to-Female 3.5mm Bullet Connectors
1	1 meter of Heat shrink wrap or electrical tape
1	1 meter 16-18 AWG Silicone Wire
1	Transmitter (Hobby King 2.4Ghz 4Ch Tx & Rx V2 - Mode 2
1	Multi-Rotor Control Board V2.1 (Atmega168PA)
1	TURNIGY BESC Programming Card

Assembly Instructions:

1. Build the Frame Body:

i) Using the 2 pieces of plywood (10cm wide x 15cm high x 0.3cm thick), cut the 2 center pieces of the frame body and drill the 0.3 cm holes using the specs on the schematic below. Save the triangular pieces of scrap wood left over, as they will be used for the landing gear in step 5.

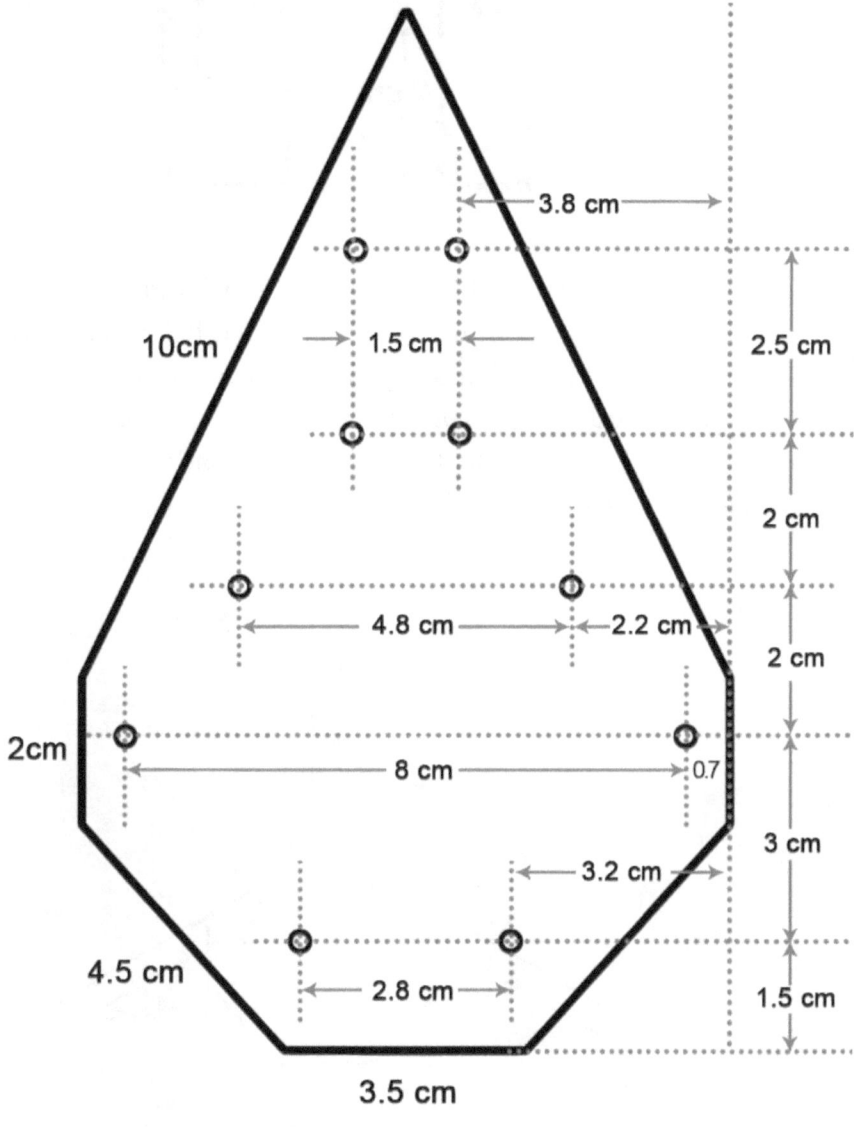

ii) On 2 of the 50 cm long pinewood arms, drill a 0.3 cm hole 2 cm from the end of the arm as shown in the picture below:

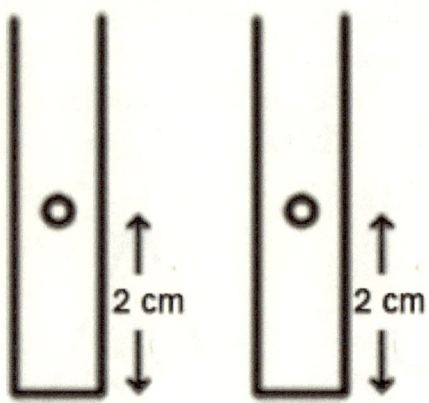

iii) Use glue to attach the 3.5cm-length piece of pinewood to the frame body as shown below. Use 10 of the "#6-32 x 1 inch screws and nuts" to assemble the arms and center piece of the frame's body. The middle arm should be stationary, and 2 of the arms are movable.

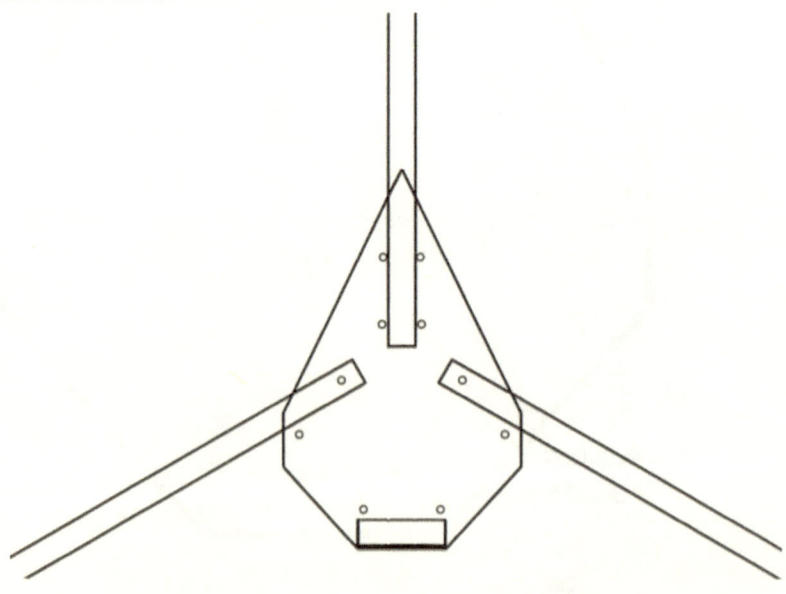

2. Assemble the Yaw Mechanism:

i) Use 1 #6-32 x 1 inch screws and glue to attach the 4cm-length piece of pinewood to the 5cm-length piece of pinewood in a "T" formation as shown in the picture below:

ii) Center the servo, attach the double-sided servo horn to the servo, then screw both ends of the double-sided servo horn of the servo horn to the "T-shaped" Yaw Mechanism as shown in the picture below:

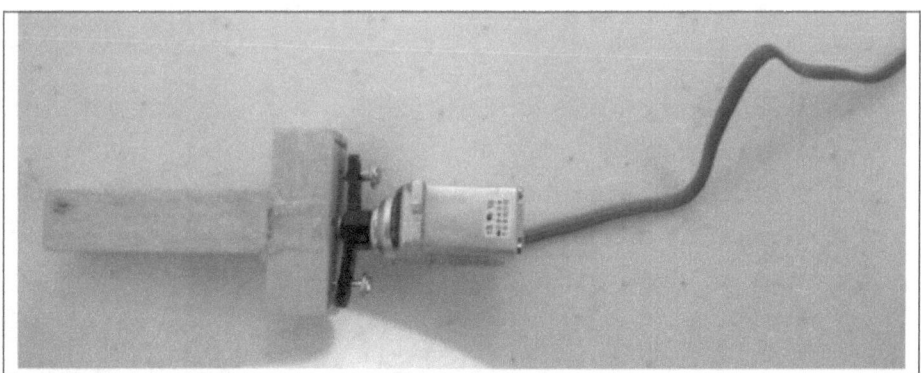

iii) Use 2 Zip ties to attach the servo to the stationary arm of the frame body as shown in the picture below:

3. Install the ESCs:

i) Solder bullet connectors to the ends of all the wires and ESCs.

ii) Use shrink wrap or electrical tape to cover the exposed metallic electrical components

iii) Zip tie the ESCs to the arms of the frame body with the wire from the ESCs to the motors pointing toward the end of the arms and the servo leads of the ESCs pointing toward the center of the body frame as shown in the picture below.

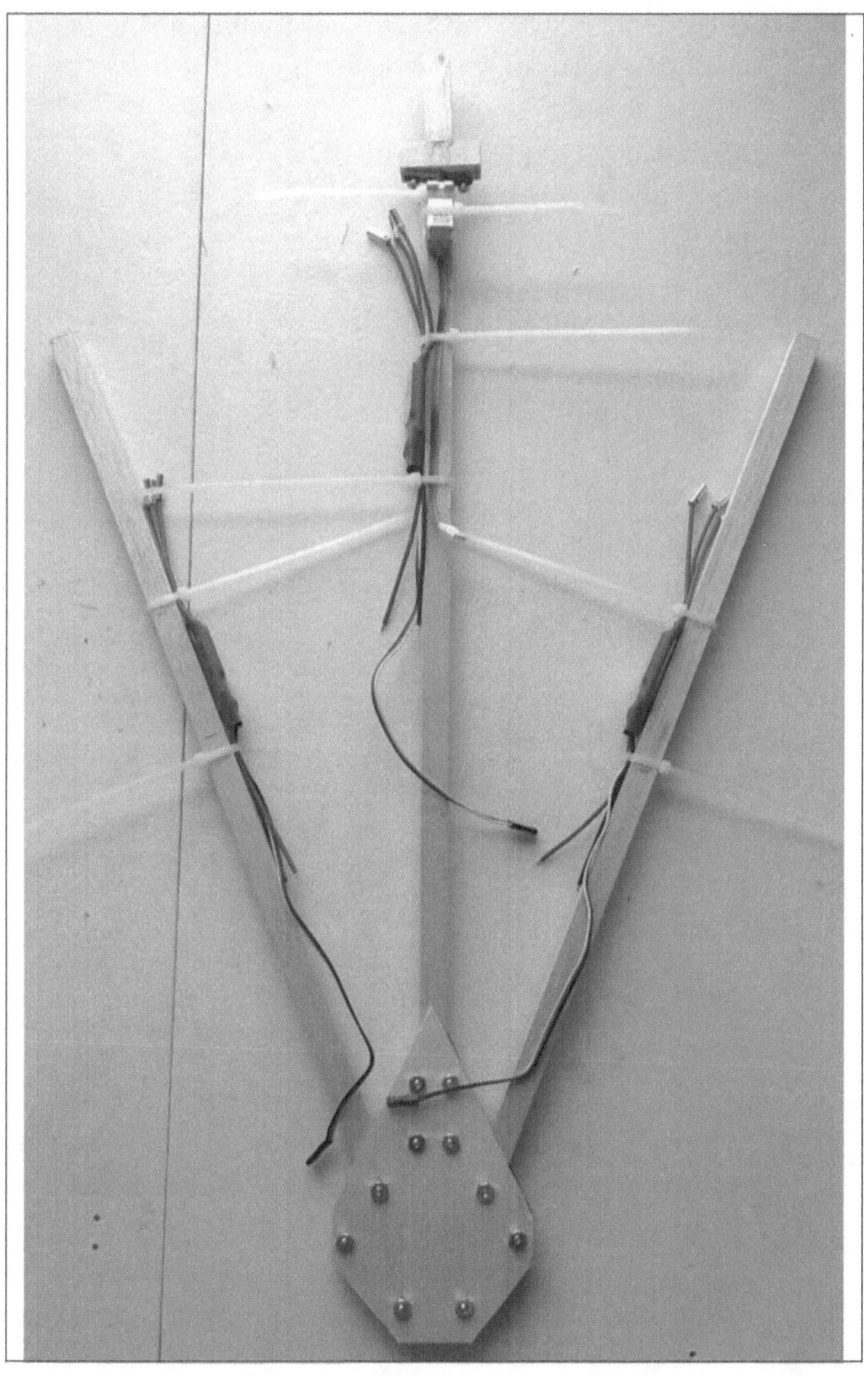

iv) Use the TURNIGY BESC Programming Card to program the ESCs with the following settings:

(1) Brake = Off

(2) Battery Type = Ni-xx

(3) Cut Off Type = Soft-Cut

(4) Cut Off Voltage = Low

(5) Start Mode = Normal

(6) Timing Mode = Middle

(7) Music = [none selected]

(8) Governor Mode = Off

4. Install the Motors:

i) Solder the bullet connectors to the motors as shown in the picture below:

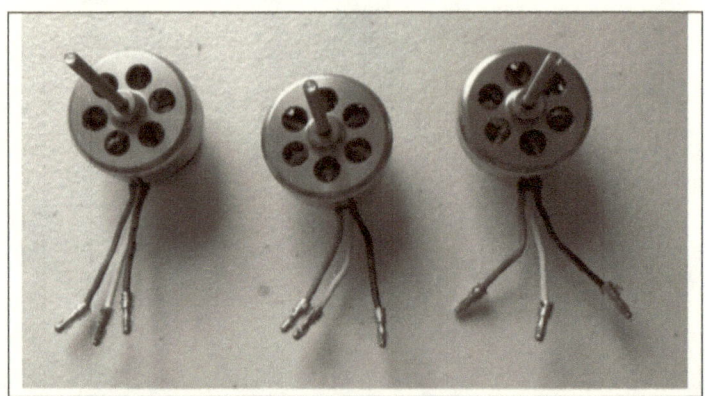

ii) Use 2 Zip ties to mount each motor to the end of each arm as shown in the picture below. Connect the Bullet Connectors of the Motor to the ESCs. If the Motor does not spin in the correct direction, then swap 2 bullet connectors with each other so that the motor spins in the opposite direction.

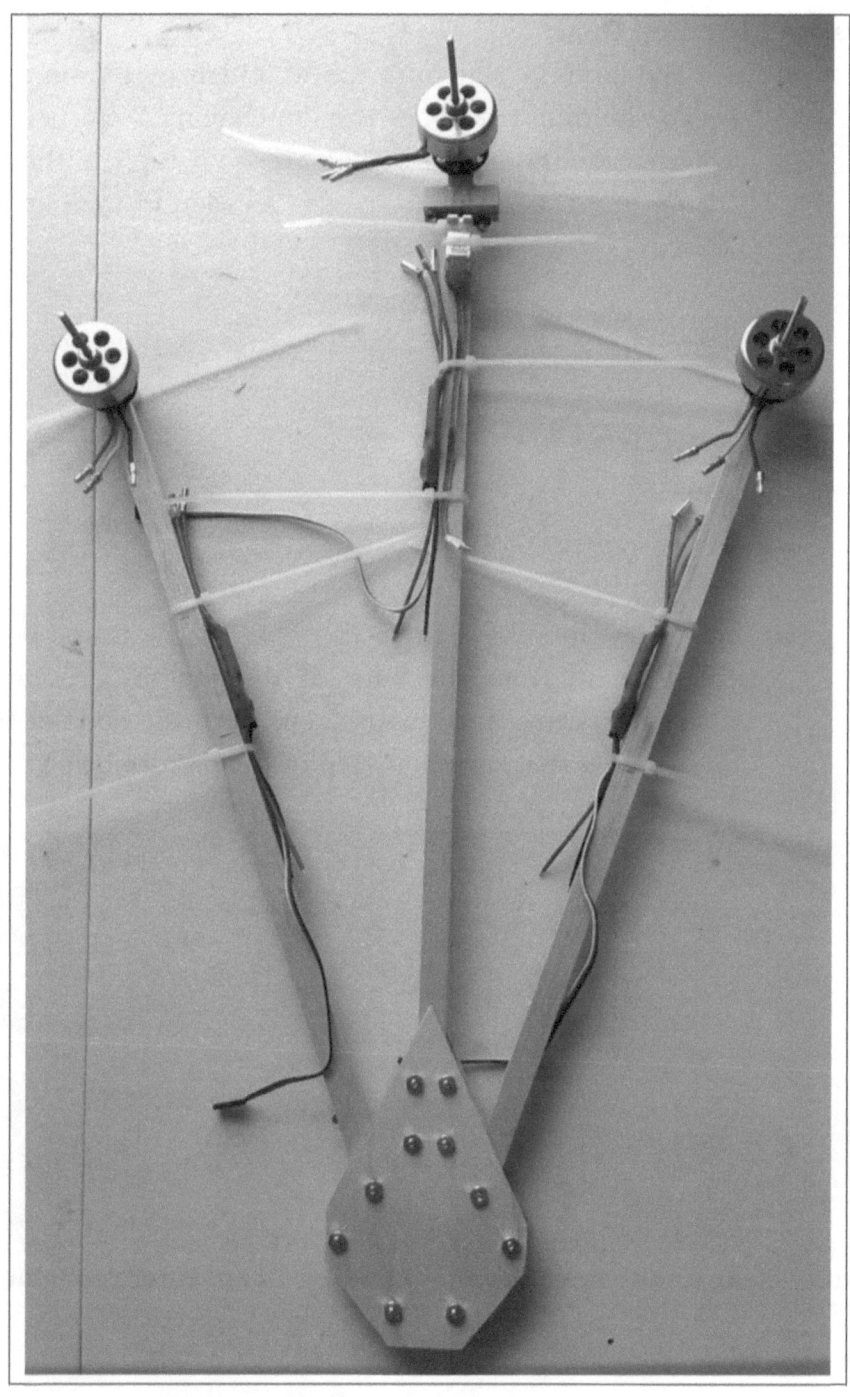

5. Install Landing Gear:

i) Cut the landing gear from the wood left over from step 1. The landing gear have the dimensions 7 cm height, 6.5 cm wide, 8cm diagonal, 1 cm from top drill holes, 1.5 cm apart, 0.75 cm from edge as seen in the picture below:

ii) Zip tie the landing gear to the arms of the frame body about 1 cm from the edge of the motors on the 2 stationary arms, and about 1 cm from the edge of the servo on the fixed arm, as seen in the picture below:

 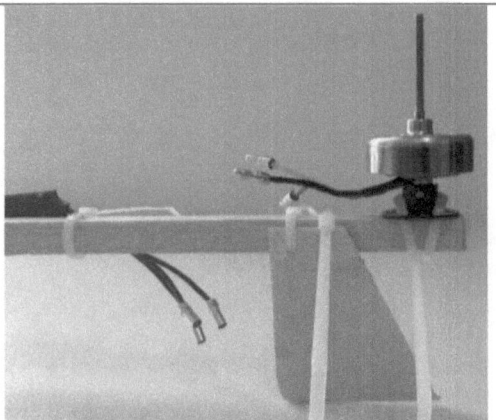

Landing Gear under fixed Arm | **Landing Gear under movable arm**

6. Install Avionics:

i) Connect the ESC wire harness to the power supply.

ii) Mount the Multi-rotor control board (gyro and receiver) to the center of the frame using double-sided sticky tape as shown in the picture below:

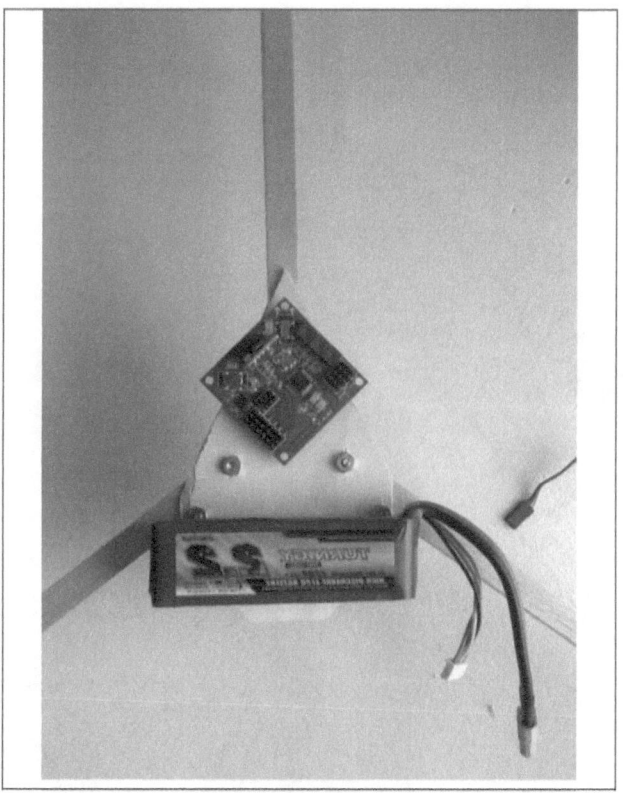

iii) Connect the ESCs and servo wires to the Multi-rotor control board.

7. Ascend:

i) Attach the Propellers, Battery, receiver, and image of Heru to the frame, and the Heru-copter Tri-copter is ready to fly and ascend.

ii) The image of Heru should be attached to the underside of the aircraft so that it can be seen by the people below on the ground.

iii) Additionally, a camera can be installed on the aircraft as an **"eye of Heru"** for aerial photography and/video.

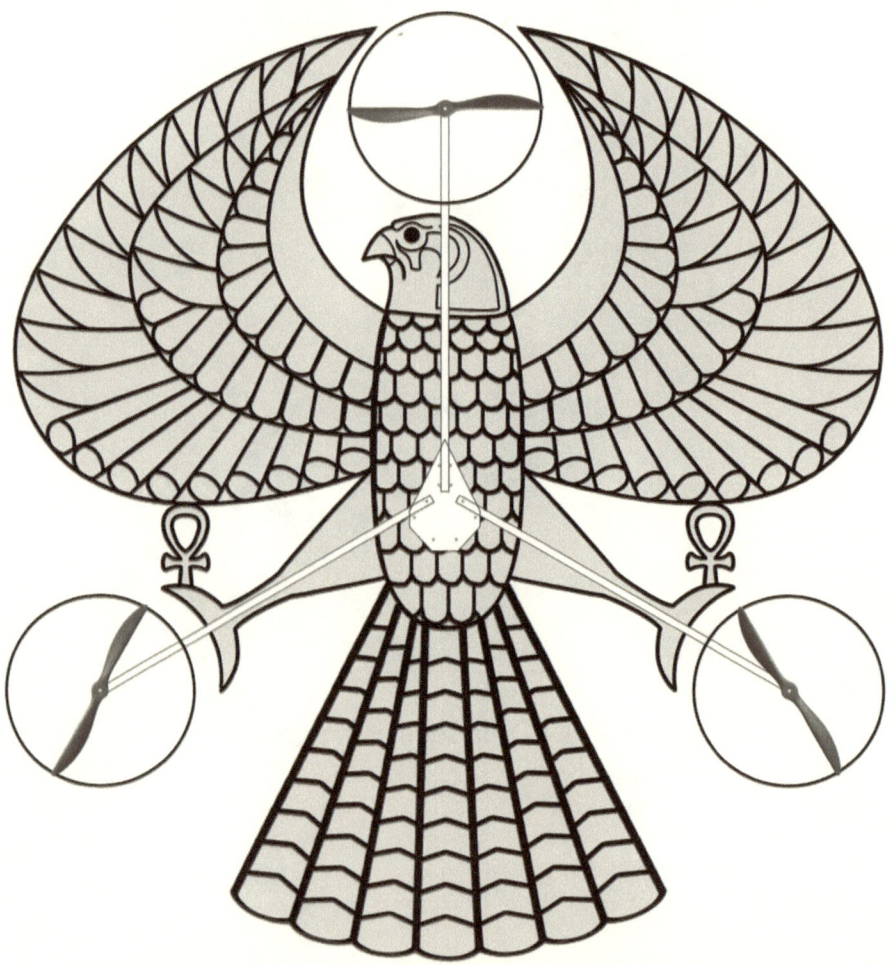

8. About the Author / Aviator

African Creation Energy can scientifically be defined as the Work, Effort, Endeavors, and Activities of African people that cause a movement or change. African Creation Energy is The Energy, Power, and Force that created African people and that African people in turn use to Create. Since African people are the Original people on the planet Earth, it follows from thermodynamics that the Creation Energy of African people is the closest creation Energy of all the people on the Planet to the **Original Creative Energies** that created the Planets, stars, and the Universe. **African Creation Energy** is **Black Power** in the scientific sense of the word "Power", and this book **Radiates** African Creation Energy to be absorbed by the **Black Body**. African Creation Energy has been called by many different names amongst many different groups of African people throughout time. African Creation Energy has been called by the names Ashe, Tumi, Dikenga, Nyama, Nzambi, Amma, Sekhem, NoopooH, and Nuwaupu just to name a few.

The conduit of "African Creation Energy" who has written and authored this book, and other books, goes by the title of **Osiadan Borebore Oboadee** from the Twi language spoken in Ghana West Africa. The Twi word "**Osiadan**" comes from the root words "Si" meaning "Build" and "adan" meaning "Building" with "O-" being a way to denote a "Master". Hence "Osiadan" literally describes a "**Master Builder**". Also note the phonetic similarities between the Twi words "Si" and "Adan" and the Ancient Egyptian words "**Sia**" (wisdom) and "**Aton**" (high noon sun). The Twi word "**Borebore**" comes from the root words "Bo" meaning "Create" and "Re" meaning "to do repetitiously", thus "Borebore" is used to describe a "**Perpetual Creator**".

The word "BoreBore" or "Bore" in Twi is also related to the Hebrew word "**Bara**" meaning "**to begin**" found in the first verse of the first chapter of the Judeo-Christian Bible, and is also related to the Yoruba word "bere" meaning "to begin". Also note the phonetic similarities between the Twi words "Bo" and "Re" and the Ancient Egyptian words "Ba" (soul) and "Re" (sun). The Twi word "**Oboadee**" comes from the root words "Bo" meaning "Create" and "Abode" meaning "Creation" with "O-" being a way to denote a "Master", hence "Oboadee" literally describes a "**Master Creator**". Oboadee is also pronounced O-Poatee in different African dialects, and is said to derive from the pronunciation of the name of the Ancient African Creation deity **PTAH**. Osiadan, Borebore, and Oboadee are three principles of **African Creation Energy**.

Osiadan is African by blood and lineage; a descendant of the **Balanta-Bassa** and **Djola-Ajamatu** tribes in present day **Guinea-Bissau** (**Ghana**-**Bassa**) West Africa. Both the Balanta and Djola tribes migrated to West Africa in Ancient times from the area which is present day **Egypt**, **Sudan**, and **Ethiopia**. Osiadan is a descendant of the Ancient **Napatan**, **Merotic**, **Kushite** Pyramid Builders, and is a Scientist, Engineer, Mathematician, Problem Solver, Analyst, Synthesizer, Artist, Craftsman, and Technologist by education, profession, and Nature. Osiadan has obtained Bachelors and Masters Degrees in the areas of Electrical Engineering, Physics, and Mathematics. Born in the African Diaspora, Osiadan made his first trip to the African continent in the year 2008. Between the years of 2009 and 2010, Osiadan Borebore Oboadee set out to develop, engineer, invent, formulate, build, construct, and create several Technologies (Applications of Knowledge) for the well being of African people worldwide and attempted to radiate the energy that motivated and inspired the development of those technologies in a three part introductory educational series which collectively was entitled "The African Liberation Science, Math, and Technology Project" **(The African Liberation S.M.A.T. Project)**. The three books that are part of African Creation Energy's "African Liberation S.M.A.T. project" are:

1. **SCIENCE:** (Knowledge/Information)
   The SCIENCE of Sciences, and The SCIENCE in Sciences
2. **MATHEMATICS:** (Understanding/Comprehension)
   9^{9^9} Supreme Mathematic African Ma'at Magic
3. **TECHNOLOGY:** (Wisdom/Application)
   P.T.A.H. Technology: Engineering Applications of African Science

The primary purpose for writing the books of the "African Liberation S.M.A.T. Project" was to motivate the Creative Energies, Minds, and Bodies of African people to go from an inert state of Theory and Speculation to an Active creative state of Development, Creation, and Productivity for the survival and well-being of African people everywhere. It is the goal of African Creation Energy's "African Liberation S.M.A.T. project" to free the minds, energies, and bodies of African people from mental captivity and physical reliance and dependence on inventions and technologies that were not developed or created by, of, and for African people.

In 2011, at the age of 30, after writing the books of the "African Liberation S.M.A.T. Project", Osiadan found it necessary to provide evidence of the **African Creation Energy** Philosophy in Action and Application by building structures, and thus embarked upon the project of building a Pyramid and authoring a text entitled "***ARCH I TET: How to Build A Pyramid***" as part of his **30 year "Djed Festival"** of renewal for all eyes to see.

In the summer of 2012, the book entitled "***9 E.T.H.E.R. R.E. Engineering***" was published which was dedicated to having readers better comprehend what they call their "**Spirit** and

Soul" by studying various aspects of **Energy** including Electricity, Thermodynamics, Hydrodynamics, Electromagnetic Radiation, and Resonant Energy, and demonstrating operative use and practical applications of these various types of Energy as a form of "**Spirituality**".

In the Winter of 2012, the book "***Khnum-Ptah to Computer: The African Initialization of Computer Science***" was published as a way to provide motivation and inspiration for Africans and people of African descent to take part in the creation and development of software, computer programming, and computer-based technologies now and on into the future, and to also show the relationships between technological concepts like Computer Science, Robotics, Virtual Reality, and Transhumanism, to various traditional African cultural and spiritual concepts like animated Statues, and Death, and Resurrection.

At the Ascension age of 33 years old, Osiadan released the book "***H.E.R.U.-copters: African Aeronautical Ascension***" on 7-7-14 to show readers how the **African Creation Energy** philosophy can be used to "**go to Heaven**" by providing motivation and inspiration to Africans and people of African descent to study and utilize the scientific fields of Aviation, Aeronautics, and Aerospace Engineering.

Just like beliefs motivate and determine actions, the religion and spirituality of a people motivates and determines their level of science, math, technology, and creativity. The 3 great realities that a true spiritual system, doctrine, or way of life, must provide people in order for that system to be complete, are:

1. Hierarchy of Needs (food, clothing, shelter, education, entertainment, social interaction)
2. Morals and Ethics

3. Answers to Existential Questions

Science, Math, and Technology are the keys to providing people with methods to obtain all 3 of these "great realities". Thus, a true spiritual system, doctrine, or way of life must include science, math, and technology in order to be complete. One of the purposes of the "**African Creation Energy**" books is to show how Ancient and Traditional African culture and spiritual systems synthesized and combined Spirituality with Science, Math, and Technology. It is the desire of **African Creation Energy** to have people of African descent return to their traditional Scientific and Mathematic Spiritual systems. One of the most prolific examples of a Scientific and Mathematic Spiritual system is the spiritual system of the Ancient Africans in Egypt, which is why the **African Creation Energy** books have utilized Ancient Egyptian symbolism and terminology to teach Scientific, Mathematic, and Technological concepts. However, Ancient Egypt was a culture, society, and empire that spanned over 3000 years, had various and multiple groups of people as dynastic rulers, and over time contained practices, teachings, and beliefs which were positive and beneficial as well as negative and destructive. Let it be known, that the utilization of Ancient Egyptian symbolism and terminology to teach Scientific, Mathematic, and Technological concepts from an African perspective in the **African Creation Energy** books does not mean we are promoting everything Ancient and everything Egyptian as good. We recognize that second to the spiritual systems of Ancient Egypt, the religion of **Islam** is a spiritual system utilized by African people which has significantly impacted and influenced Scientific, Mathematic, and Technological development in the world; we also recognize that both of these spiritual systems have had their negative shortcomings as well. With the recognition that spiritual systems are created by human beings, for human beings, just like technology, then we must also recognize when that technology, or spiritual system, has become outdated, corrupted, and started to malfunction. When a Spiritual

System, or technology, becomes outdated and starts to malfunction, it is the responsibility of the creators of that technology to **Upgrade** and **Update** the technology by using reason and discernment to select the best working parts of the technology, and throw away the corrupted malfunctioning parts of the technology. Likewise, a similar upgrade and update is needed at this point in time for the spiritual systems of African people worldwide. It is the goal of **African Creation Energy** to provide the upgrade and update to the African Scientific Spiritual System.

One of the more contemporary systems that people are familiar with, which combined spirituality and science, is called **Alchemy**, coming from the Arabic word *Al-Khemi*, meaning **"The Black"**. Alchemical principles have been used in the development of the "African Creation Energy" books as a way to bring about the desired change and show how science and spirituality can coexist. The chosen term **"African Creation Energy"** has hidden Alchemical meaning. The word **"Africa"** comes from the Afro-asiatic word **"Afar"** meaning **"dust"** which represents the **"Earth"**. The word "Africa" is also related to the Afro-Asiatic trilateral root F-R-Q, as in the word Furqan, meaning *"the criteria for discernment or separation"*, or the name Farooq meaning *"one who knows truth from falsehood"*. The etymology of the word **"Creation"** comes from the word **"Crescent"** which represents the **"Moon"**, and the word **"Energy"** represents the **"Sun"**, which is a primary source of energy to our planet. Therefore, one of the hidden meanings in the coined term **"African Creation Energy"** represents the **"Earth**, **Moon**, and **Sun"** cosmic forces also known as **"Ptah, Aah,** and **Re"** or **"Re Ah Ptah"** or **"Space**, **Matter**, and **Time"** in African Cosmology. Moreover, the abbreviation of **"African Creation Energy"** is **A.C.E.** which spells the word "Ace" which has etymological meanings of "a unit, whole, one, first, one who excels, and

Primary" and indeed African Creation Energy represents the excellent Primary and **Original Creative Forces** in Nature.

The letters used to abbreviate "African Creation Energy", A.C.E., not only spell the English word "Ace" meaning **"First, Primary, or Original"**, but also represent the 3 fundamental geometric shapes of the **Triangle** represented by the letter **"A"**, the **Circle** represented by the letter **"C"**, and the **Square** represented by the letter **"E"**,

A.C.E.
African Creation Energy
Alchemical "Squared Circle"

which when combined form the Ancient **Alchemical symbol** of the **"Squared Circle"**. All of the books authored by Osiadan Borebore Oboadee and African Creation Energy have been released on specific strategic dates, and each of the previously released books represents a different element necessary for transformation in Ancient African Alchemy. The Alchemical classical elements of Air, Water, Fire, and Earth are represented by the African Creation Energy books entitled *"The SCIENCE of Sciences, and The SCIENCE in Sciences"*, "Supreme Mathematic African Ma'at Magic", "P.T.A.H. Technology", and *"ARCH I TECT"* respectively. The quintessential element, or 5th element, of Ether or Spirit, is represented by the African Creation Energy book entitled *"9 E.T.H.E.R. R.E. Engineering"*. The African Creation Energy book entitled *"Khnum-Ptah to Computer"* represents death, resurrection, and after-life, and the book *"H.E.R.U.-copters"* represents "ascension into heaven". These **7 A.C.E.** books are the *"7 Seals"* or *"7 Thunders"* which provide a foundation for a scientific-based African spiritual system which can in turn be utilized to create a reality for your liberation. The "Science of Creating" is the information that is most needed by people of African descent right now and on into the future.

If a Technology or ANY creation or invention is being used by a group of people, and that Group of People is dependent on that technology for Survival and Well Being, but that Group of people is not in control of the Creation or Production of that Technology, then that Group of People are Literally **SLAVES** to the Creators and the Producers of the Technology. As Africans and people of African descent look to the Future and we see the Rapid Advances of Technology, but we do not see ourselves as the Developers, Creators, Inventors, Designers, or Producers of the Technology, then a sense of Hopelessness, despair, desperation, anger, helplessness, Powerlessness, and feelings of oppression arise. **African Creation Energy** has shown that we as African people have always been **Creators**, **Fashioners**, and **Makers** of Technology and we do indeed have a place as the Developers, Inventors, and Designers of all the highly advanced technologies which will shape the Future. In the near Future, the ability to be **conscious**, or **aware**, or **knowledgeable** about something will become trivial, and thus the next level of consciousness, will become **ACTIVENESS** and having the ability to use and put all of the knowledge and information which one is conscious and aware of, into practical application. No longer will "Knowing" be a big deal because almost everyone will "KNOW", the next level will be **"how can what is known be used and applied"**. Knowledge is a natural resource. It is the practical application and utilization of all natural resources, including knowledge, which determines economic prosperity. Following the plethora of information presented by the many great African Scholars (who have affectionately been labeled **"MASTER TEACHERS"**) who have came to improve the conditions of African people; it is the goal of **African Creation Energy** to be the catalyst in the synthesis, unification, and practical application of the information presented by the great Master Teachers. Thus, it is the aspiration of "African Creation Energy" to be, and breed, **"Master Technicians"** who TEaCH through Action and Application.

African Creation Energy (A.C.E.) is dedicated to **African Centered Education** (A.C.E.) and one of the goals of African Creation Energy is to introduce the utilization of African terminology and symbolism in Science, Technology, Engineering, and Math (S.T.E.M.) education. African Creation Energy is the **ROOT of S.T.E.M.** (Science, Technology, Engineering, and Math). In military terms, a "Flying Ace" is an aircraft pilot who has shot down at least 7 enemy aircraft while in battle. The 7 African Creation Energy books have metaphorically "shot down" any erroneous notions regarding the lack of involvement of African people in the S.T.E.M. fields, and thus African Creation Energy can be considered a metaphorical **"Flying A.C.E."**

Scientist, Mathematician, Electrical Engineer, Architect, Computer Programmer, Roboticist, and Aviator, Osiadan is a ***"Jack of all trades and a Master of Nun"*** (Nun is the Ancient African Egyptian name for the Original Creative Forces in Nature). **"Africa"** (F-R-Q) is wherever we are in our **right mind**; and this includes on other continents or other planets. It is a fact that our sun and planet Earth

The Aviator, Prophessor A.C.E.

will not always be in a habitable condition, and therefore the science of space exploration is a science necessary for the Survival and wellbeing of African people in the future. **"Afro-futurology"** is an "African-centered" scientific and mathematic-based "prophecy" methodology to *"foretell the future"*. In the future, as Africans began to explore the cosmos, we will go from being "Afro-Centric" to being **"Astro-Centric"**, and **African Creation Energy** is the Escalator, the Elevator, and the Aviator which rises you up and takes you higher.

African Creation Energy Books

KHNUM-PTAH To COMPUTER:
The African Initialization of
Computer Science
Release Date: 12-12-12

HERU-COPTERS:
African Aeronautical Ascension
Release Date: 07-07-14

ARCH I TECT:
How to Build A Pyramid
Release Date: 11-11-11

9 E.T.H.E.R. R.E. Engineering
Release Date: 06-26-12

The SCIENCE of
Sciences and
The SCIENCE in
Sciences
Release Date: 10-10-10

9^{99} Supreme
Mathematic,
African Ma'at Magic
Release Date: 09-09-09

**P.T.A.H.
Technology:**
Engineering
Applications of
African Sciences
Release Date: 05-04-10

9. Appendix

REFERENCES

1. "9^{9^9} Supreme Mathematics African Ma'at Magic" By African Creation Energy

2. "9 E.T.H.E.R. R.E. Engineering" By African Creation Energy

3. "Ancient Egypt, The Light of the World" by Gerald Massey 1907

4. "ARCH I TECT: How to Build A Pyramid" By African Creation Energy

5. "Black wings" by Lieut. William J. Powell, 1899-1942
6. "Flying Saucer Technology" by Bill Rose
7. "Make Your Own Tricopter" http://www.instructables.com/id/TRICOPTER/?ALLSTEPS
8. National Archives: "Project 1794 Final Development Summary Report 2 April - 30 May 1956, USAF Contract No. AF33(600) 30161 I.D. No. 56-RDZ-19954 AVRO AIRCRAFT LIMITED
9. "P.T.A.H. Technology: Engineering Applications of African Sciences" By African Creation Energy

10. "Reinterpretations of the Ankh Symbol Part 2" by Asar Imhotep
11. "Religulous", Bill Maher 2008
12. San Diego Air and Space Museum http://www.sandiegoairandspace.org/exhibits/african_american_exhibit/timeline.php
13. "Stolen Legacy" by George G.M. James

14. "The Ancient Gods Speak: A guide to Egyptian religion" by Edmund Meltzer
15. "The Prophet: On Freedom" by Kahlil Gibran

16. "The SCIENCE of Sciences and the SCIENCE in Sciences" By African Creation Energy

17. The Tricopter V2.5 - RC Explorer http://rcexplorer.se/projects/2011/09/the-tricopter-v2-5/
18. "Zeitgeist: The Movie", 2007

PHOTO CREDITS

- Emory Conrad Malick, Source: http://www.emoryconradmalick.com
- "F-35B Joint Strike Fighter's thrust vectoring nozzle and lift fan", by Tosaka, 27 July 2008
- Types of Gas Turbines, Source: https://www.grc.nasa.gov/www/k-12/airplane/trbtyp.html

www.ingramcontent.com/pod-product-compliance
Lightning Source LLC
Chambersburg PA
CBHW022058170526
45157CB00004B/1396